DK

WHY?

WEATHER

Brilliant answers to baffling questions

DK

WHY?
WEATHER

Steve Setford

DK Penguin Random House

Author Steve Setford
Consultant Professor Adam Scaife
Illustrator Dan Crisp

DK LONDON
Editors Abi Maxwell, John Hort
Senior Art Editor Ann Cannings
Managing Editor Gemma Farr
Managing Art Editor Diane Peyton Jones
Production Editor Dragana Puvacic
Production Controller Rebecca Parton
Jacket Designer Ann Cannings
Art Director Mabel Chan

DK DELHI
Assistant Editor Syed Tuba Javed
Senior Art Editor Kanika Kalra
Project Art Editor Bhagyashree Nayak
Art Editor Nishtha Gupta
Senior Jacket Designer Rashika Kachroo
DTP Designers Ashok Kumar, Anita Yadav
Picture Researcher Ridhima Sikka
Deputy Managing Editor Roohi Sehgal
Managing Editor Monica Saigal
Managing Art Editor Ivy Sengupta
Delhi Creative Head Malavika Talukder

First published in Great Britain in 2024 by
Dorling Kindersley Limited
DK, One Embassy Gardens, 8 Viaduct Gardens,
London, SW11 7BW

The authorised representative in the EEA is
Dorling Kindersley Verlag GmbH. Arnulfstr. 124,
80636 Munich, Germany

A CIP catalogue record for this book
is available from the British Library.
ISBN: 978-0-2416-8669-0

Printed and bound in India

Contents

Making weather

Weather patterns

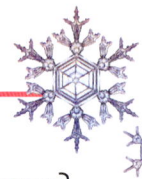

Find out how guanacos survive the thin air in high mountains on page 101.

Weather around the world

Extreme weather

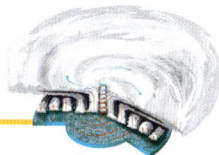

Living with weather

Climate change

QUICK QUIZ

Test your knowledge! Look out for the "Quick quiz" box throughout this book to see how much you've learned. Turn to pages 132–133 for the answers.

Making weather

The powerhouse that drives our weather is the Sun. Even though it lies about 150 million km (93 million miles) away, the Sun stirs up Earth's atmosphere with its energy. Weather is the result of this stirring. Weather isn't limited to Earth – there is weather on other planets, too!

Why do we have weather?

We have weather because the air and water in Earth's atmosphere are continually on the move, set in motion by energy from the Sun. These movements create wind, rain, and all other types of weather.

Sun's heat

The Sun's heat warms air in the lower atmosphere most strongly near the Equator, where the surface of the Earth faces the Sun. Warm air rises, cool air sinks, and this movement starts winds blowing.

What is air made of?

Nitrogen and oxygen are the two main gases in air, but there are tiny amounts of other gases, too. In the lower atmosphere, air also contains water vapour – water in its gas form.

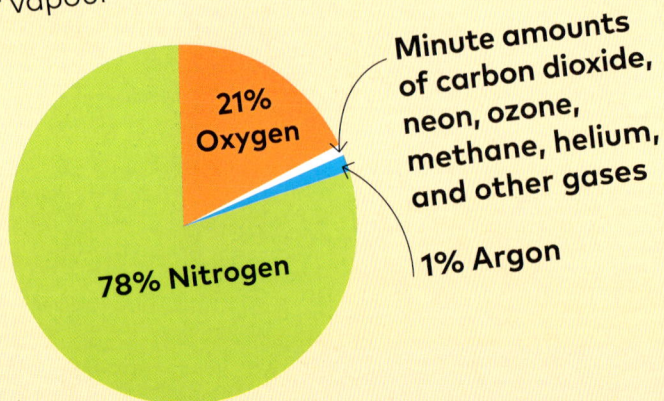

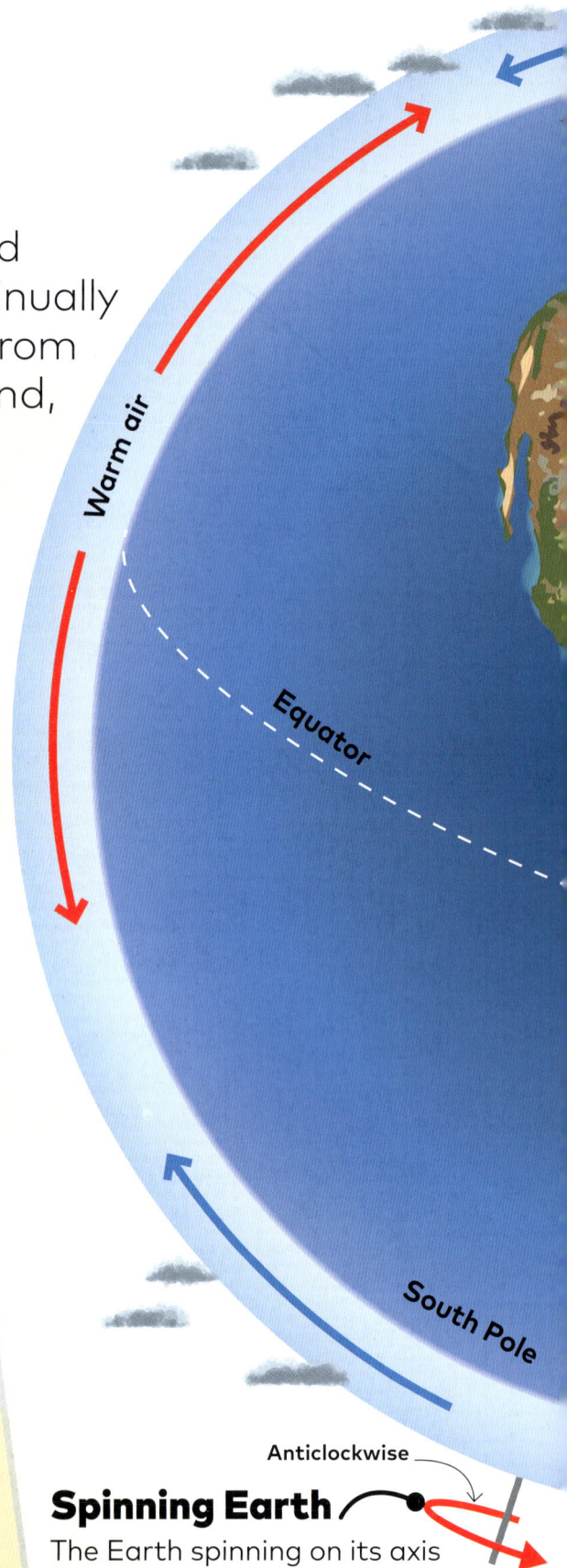

21% Oxygen

Minute amounts of carbon dioxide, neon, ozone, methane, helium, and other gases

1% Argon

78% Nitrogen

Warm air

Equator

South Pole

Anticlockwise

Spinning Earth

The Earth spinning on its axis also affects the direction in which the winds blow, especially away from the Equator.

Axis

Cool air

North Pole

Warm and cool air move around

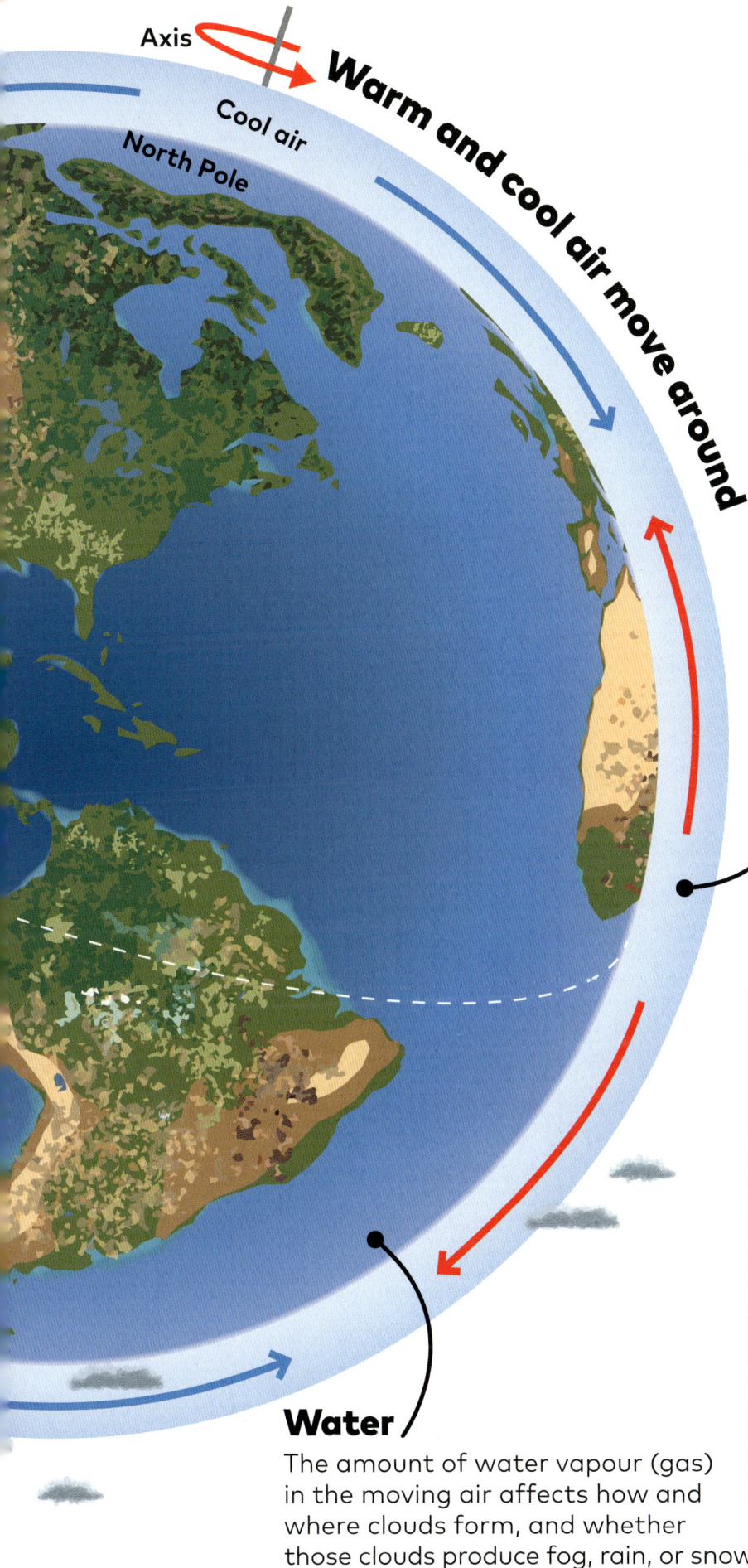

Is weather the same as climate?

Climate is the usual pattern of weather somewhere, over many years. Weather is a short-term event, such as a storm or a hot day. A place with a dry, sunny climate may still have some days when the weather is overcast and rainy.

Atmosphere

The Earth is surrounded by a gaseous blanket called the atmosphere. It is made of a mixture of gases called air.

Water

The amount of water vapour (gas) in the moving air affects how and where clouds form, and whether those clouds produce fog, rain, or snow.

TRUE OR FALSE?

1. Air is mostly oxygen.

2. Climate is the long-term weather.

3. The air in Earth's atmosphere is always on the move.

4. Warm air sinks and cool air rises.

See pages 132–133 for the answers

What does the atmosphere do?

Earth's atmosphere keeps the planet warm and provides air for living things to breathe. It also helps to block harmful rays from the Sun, and protects us from space rocks. The air in the atmosphere has weight, so it presses down on Earth's surface and against anything it touches.

THE OZONE LAYER HELPS TO BLOCK **ULTRAVIOLET (UV) RAYS** FROM THE SUN, WHICH CAN HARM LIVING THINGS.

Our planet's armour

The atmosphere has five distinct layers. As you travel up from the ground towards space, the air in each layer becomes less dense (thinner) than the air in the layer below.

Meteors

Weather balloon

Aeroplane

What is air pressure?

Air pressure is a measure of how much the air is pushing down on things. It is high at sea level because the weight of the whole atmosphere is pushing down. It is lower on top of a high mountain, where there is less atmosphere above.

Exosphere
The exosphere reaches to thousands of kilometres above Earth's surface. At its top, it merges with space.

Hubble Space Telescope

Thermosphere
The thermosphere is hundreds of kilometres deep. The International Space Station orbits Earth in this layer.

International Space Station

Mesosphere
The mesosphere is 30 km (20 miles) deep, and where small space rocks burn up. You can see them as streaks of light, called meteors, or shooting stars.

Aurora

Stratosphere
The stratosphere is about 35 km (22 miles) deep. It contains a layer of ozone gas.

Troposphere
Weather happens in the troposphere. This layer is 8 km (5 miles) deep over the poles and 18 km (11 miles) deep over the Equator.

QUICK QUIZ

1. In which layer of the atmosphere does weather happen?
   a) Stratosphere
   b) Thermosphere
   c) Troposphere

2. Where is the ozone layer?
   a) Stratosphere
   b) Thermosphere
   c) Troposphere

See pages 132–133 for the answers

Big chill

Temperatures would quickly tumble. Without the Sun's energy, weather as we know it would cease. Earth's surface water, including the top layers of the ocean, would start to freeze over.

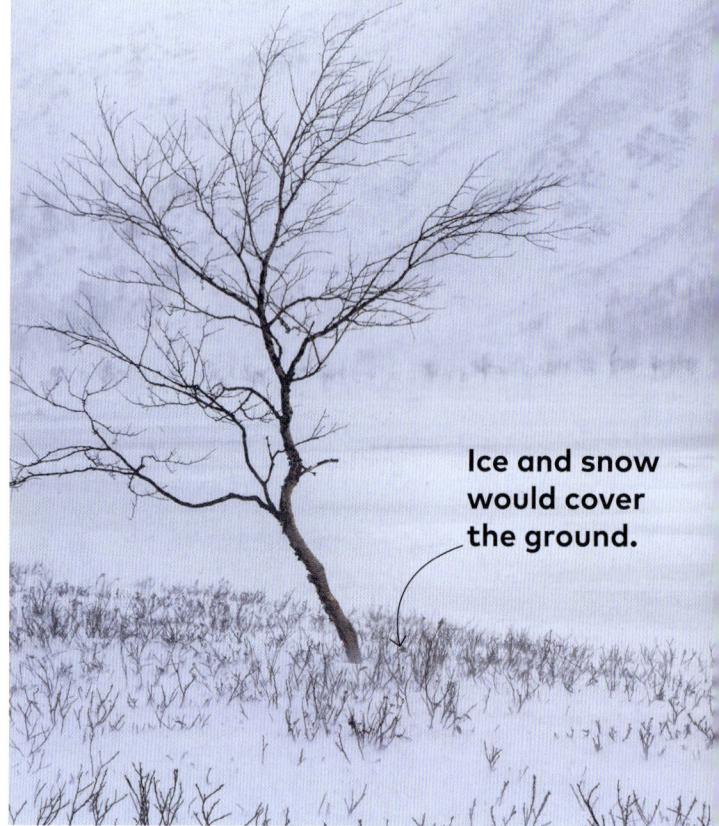

Ice and snow would cover the ground.

Lights out

We wouldn't notice anything at first, because it takes about 8 minutes for light released by the Sun to reach the Earth. After that, darkness would fall.

What if the Sun stopped shining?

If the Sun stopped shining, life on Earth would perish. The planet would be plunged into darkness and finally become an ice-coated ball of rock. Luckily, the Sun can't suddenly "switch off". It won't last forever, but don't worry – the Sun has a very, very long life ahead of it!

AT ABOUT 4.5 BILLION YEARS OLD, THE SUN IS **ROUGHLY HALFWAY** THROUGH ITS LIFE.

No food, no life

Plants use the energy in sunlight to make their food. Without sunlight, plants would die. So, too, would all plant-eating animals, and the animals that prey on the plant-eaters.

Our planet would become a ball of ice.

Snowball Earth

Eventually, the ocean depths would turn to ice, too. Even the air would condense and freeze onto Earth's surface. Earth would end up mega-chilly, perhaps −240°C (−400°F).

How long will the Sun last?

The Sun will stay like it is for several billion years. Then it will swell up and become a red giant star. Finally, it will puff off its outer layers, leaving just a fading, cooler core, called a white dwarf.

A red giant reaches up to a billion kilometres (620 million miles) in diameter.

QUICK QUIZ

1. About how long does it take for light from the Sun to reach the Earth?

2. Why do plants need sunlight?

3. Billions of years in the future, will the Sun shrink or swell up?

See pages 132–133 for the answers

What is the greenhouse effect?

Some of the gases in Earth's atmosphere act like the glass roof of a greenhouse – they trap heat radiating from the Earth, warming the planet. This is called the greenhouse effect. Without the greenhouse effect, most of the heat would escape, and Earth would be too cold to sustain life.

VENUS HAS A POWERFUL GREENHOUSE EFFECT, SO THE AVERAGE SURFACE TEMPERATURE IS ABOVE 500°C (930°F).

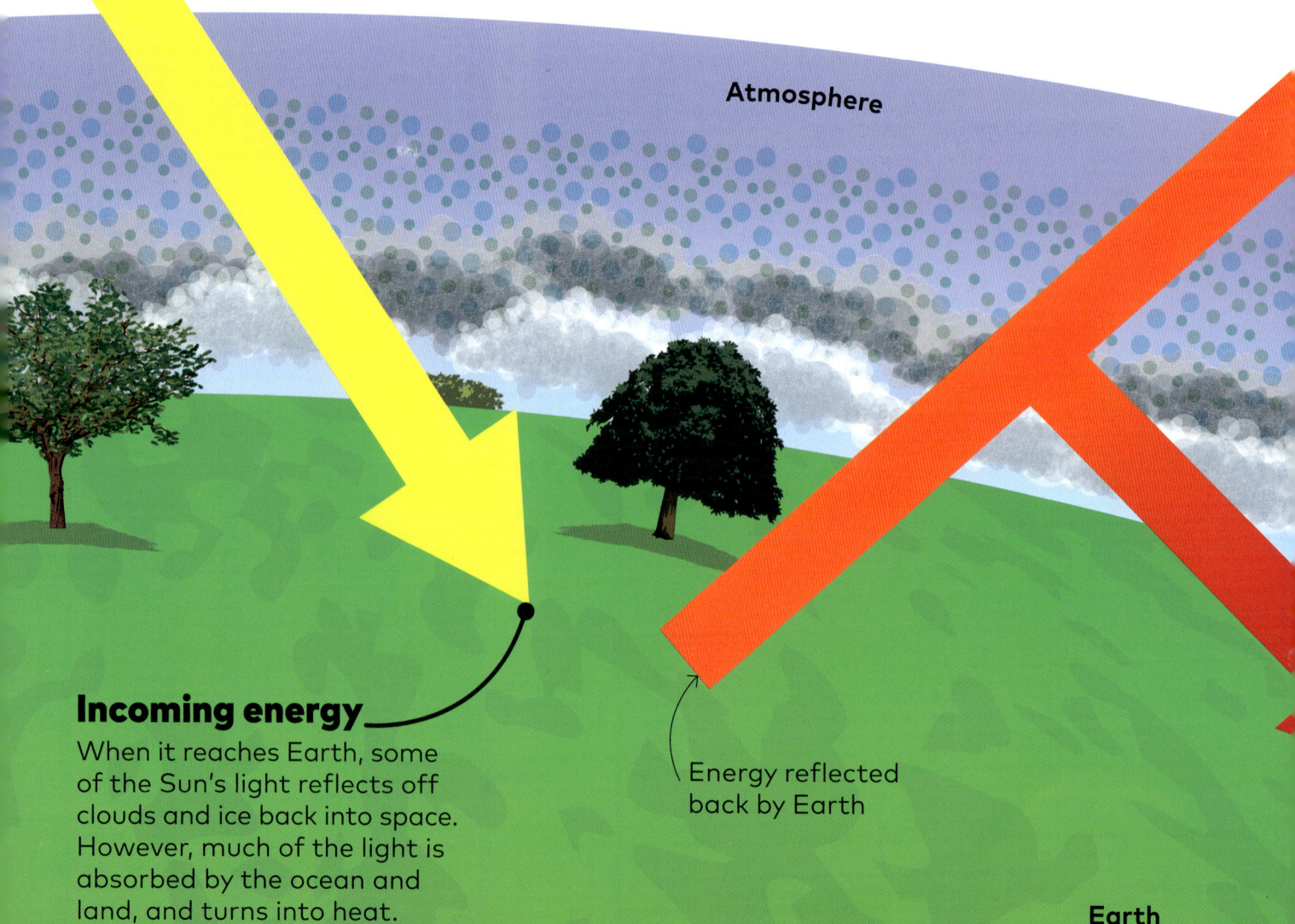

Atmosphere

Incoming energy

When it reaches Earth, some of the Sun's light reflects off clouds and ice back into space. However, much of the light is absorbed by the ocean and land, and turns into heat.

Energy reflected back by Earth

Earth

How is human activity increasing the greenhouse effect?

Burning fossil fuels (coal, oil, and natural gas), farming, and other human activities such as transportation, are adding extra greenhouse gases to the atmosphere. More heat is getting trapped and temperatures are rising around the world. This is called global warming.

1. Carbon dioxide in the atmosphere absorbs the Earth's heat energy.

2. The natural greenhouse effect makes life on Earth possible.

3. Burning fossil fuels removes greenhouse gases from the atmosphere.

See pages 132–133 for the answers

Escaping heat
The warmed Earth releases heat energy back into the atmosphere. Some of this escapes into space.

Greenhouse gases

Carbon dioxide

Carbon O C O

Methane
C H
Hydrogen

Water vapour
Oxygen H O H

Nitrous oxide
Nitrogen N N O

Trapped heat
Greenhouse gases in Earth's atmosphere trap the rest of the heat energy. They stop it from escaping and keep the planet warm.

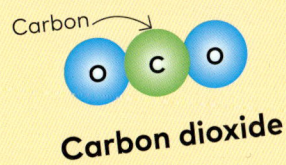

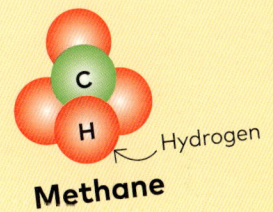

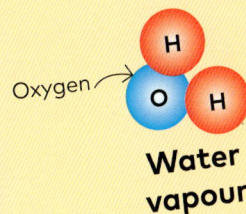

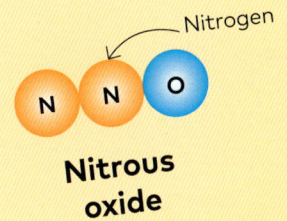

What are the main greenhouse gases?
The main greenhouse gases are carbon dioxide, methane, water vapour (water in gas form), and nitrous oxide. They let light pass through the atmosphere, but they stop the Sun's heat from escaping Earth.

Has weather always existed?

No, weather hasn't always existed on Earth. This is because when Earth formed, there was no real atmosphere in which weather could happen. An atmosphere did gradually form, but we know very little about what the weather on the young Earth was like.

ABOUT HALF OF EARTH'S OXYGEN COMES FROM **LIVING THINGS IN THE SEA,** AND HALF FROM **LAND PLANTS.**

Molten rock ball

About 4.5 billion years ago, Earth formed from a hot mixture of gases and solid particles. The young Earth had a surface of molten rock, and almost no atmosphere.

Early atmosphere

Earth's surface cooled and solidified. An atmosphere formed as gases gushed from volcanoes. It probably included lots of carbon dioxide, plus water vapour, and some other gases – but no oxygen.

What is photosynthesis?

Photosynthesis is the way plants, algae, and certain bacteria use the energy in sunlight to make food from water and carbon dioxide. As they do so, they release oxygen into the air. The plant cells contain a pigment called chlorophyll, which absorbs the sunlight and reflects green light back, making the plant appear green.

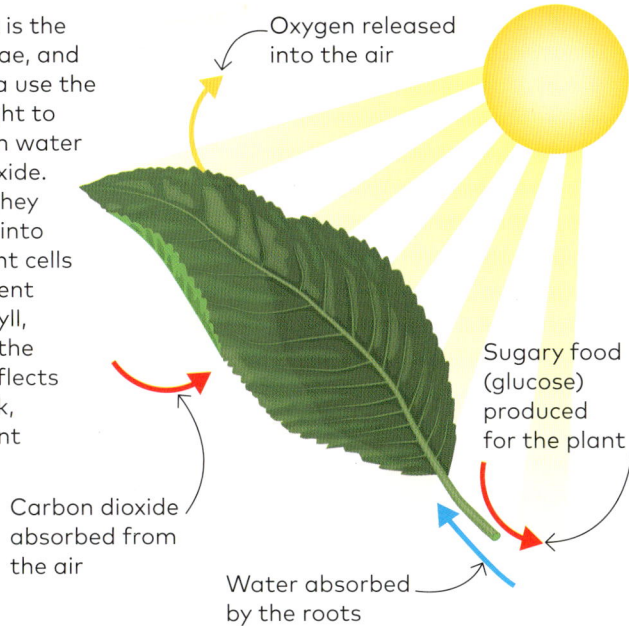

Oxygen released into the air

Sugary food (glucose) produced for the plant

Carbon dioxide absorbed from the air

Water absorbed by the roots

QUICK QUIZ

1. What were the first living things to start adding oxygen to the atmosphere?

2. Where did the gases that made up the early atmosphere come from?

3. For how long has the mix of gases in the atmosphere been like it is today?

See pages 132–133 for the answers

Rain, oceans, and oxygen

The water vapour condensed into clouds. Rain fell and created the oceans, which is where life began. Ocean-dwelling microbes, called cyanobacteria, started to release oxygen into the atmosphere through photosynthesis.

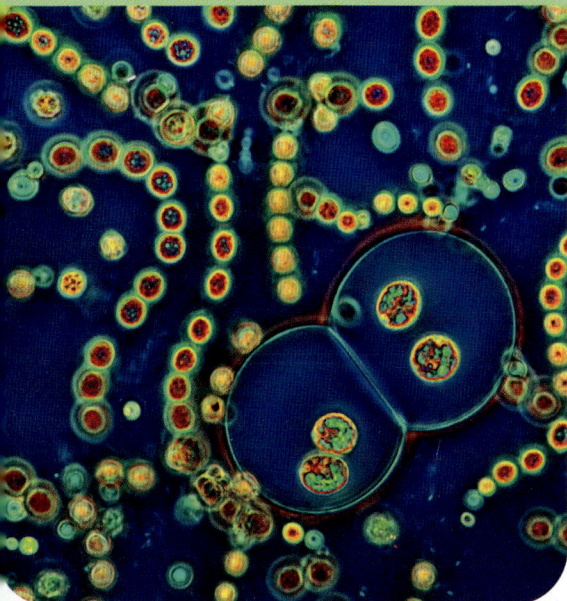

Today's atmosphere

As plants spread over the land, they too added oxygen to the atmosphere. For the last 200 million years, the proportions of the different gases in the atmosphere have stayed roughly the same.

March

In March, it is autumn south of the Equator, and spring north of the Equator. The Sun shines equally on the two hemispheres.

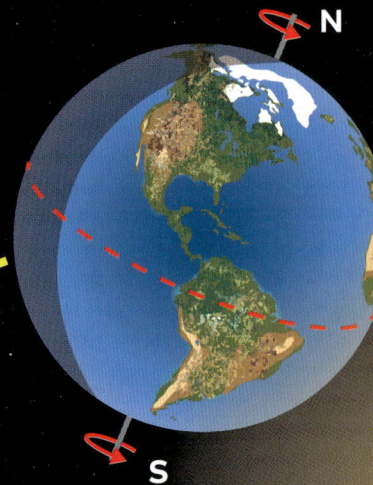

PLACES NEAR THE EQUATOR GET **EQUAL SUNLIGHT ALL YEAR ROUND**, SO THEY ONLY HAVE TWO SEASONS: WET AND DRY.

Direction of Earth's rotation

June

In June, the North Pole tilts towards the Sun, bringing summer to the northern hemisphere. It is winter in the southern hemisphere.

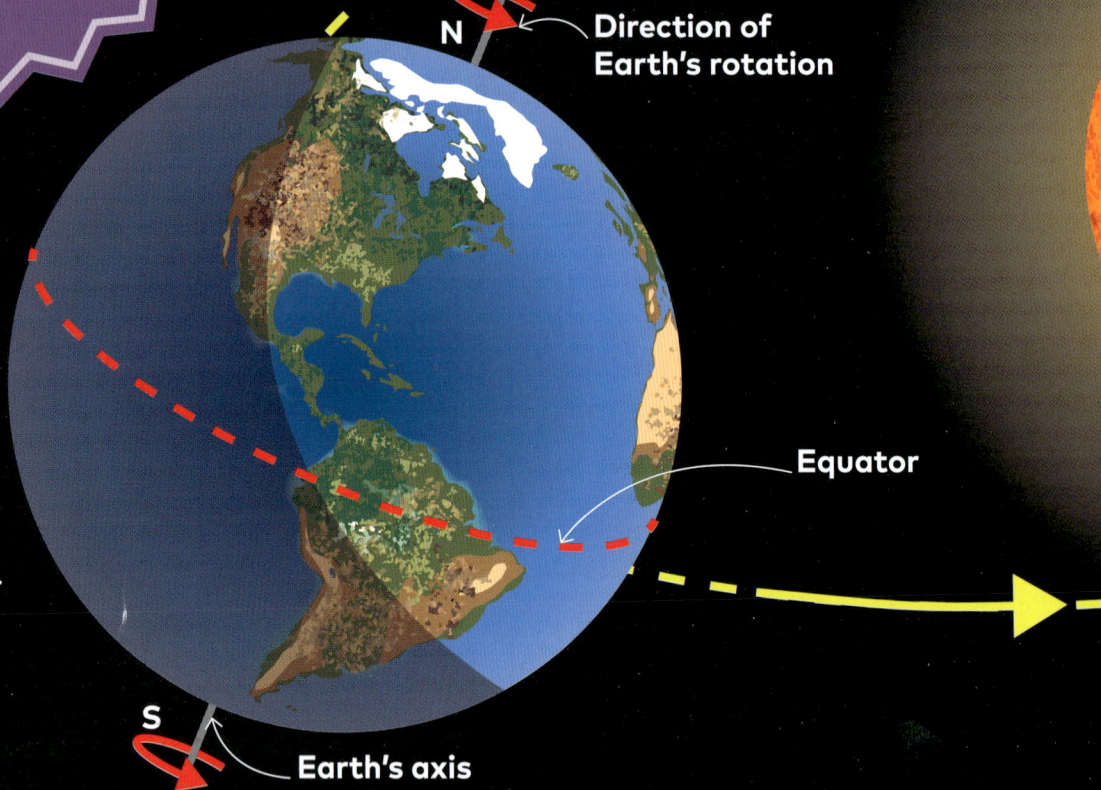

Equator

Earth's axis

Why do seasons change?

Seasons change because Earth is tilted on its axis – the imaginary line around which our planet spins as it orbits the Sun. Throughout the year, different parts of the Earth receive more sunlight as they point towards the Sun, then less sunlight as they point away from it. As sunlight levels change, so too do the seasons.

December

In December, when the South Pole tilts towards the Sun, it is summer in the southern hemisphere. In the northern hemisphere, it is winter.

N

Northern hemisphere

S

Southern hemisphere

Earth's orbit around the Sun

Sun

N

S

September

In September, the Sun shines equally on both hemispheres. It is spring south of the Equator, and autumn north of the Equator.

Is Earth's orbit a perfect circle?

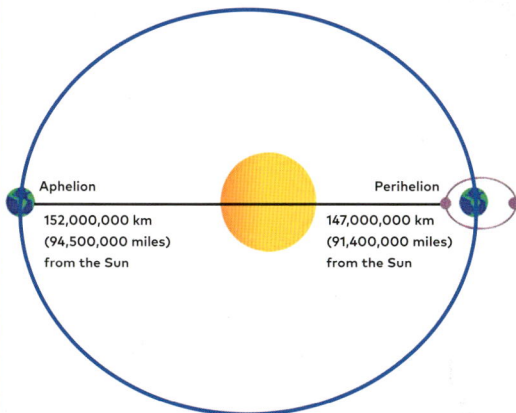

Aphelion
152,000,000 km
(94,500,000 miles)
from the Sun

Perihelion
147,000,000 km
(91,400,000 miles)
from the Sun

No, Earth's orbit is slightly oval-shaped, so our distance from the Sun changes during the year. We call the closest point perihelion, and the furthest point aphelion. Compared with how far away the Sun is, the difference between perihelion and aphelion is relatively small, so it does not affect the seasons.

TRUE OR FALSE?

1. Every part of Earth gets the same amount of sunlight all year round.

2. Earth's closest point to the Sun is the aphelion.

3. Earth's orbit is slightly oval-shaped.

See pages 132–133 for the answers

Temperature

When talking about the weather, temperature describes how warm or cold the air is. Warm air expands, and cold air contracts.

Air pressure

Air pressure is the weight of the air above your head. This is why the pressure is low at the top of a mountain, as there's less air above you.

Wind

Winds blow around low- and high-pressure centres. Changes in pressure change the winds. They can blow rain clouds in – then blow them out again.

Why does the weather change?

Weather has six main ingredients: temperature, air pressure, wind, humidity (the amount of moisture in the air), precipitation (rain and snow), and cloudiness. If any of these alters, the weather changes. All weather is connected: a change of weather in one place sends ripples of change through the weather elsewhere.

FIVE-DAY FORECASTS ARE ACCURATE **90 PER CENT** OF THE TIME. 10-DAY FORECASTS ARE ACCURATE ABOUT **HALF THE TIME.**

Humidity

Warm air can hold more moisture than cold air, so it has a higher humidity. This means more chance of precipitation.

Precipitation

Rain, snow, sleet, and hail are all types of precipitation. Precipitation usually happens when warm, humid (moist) air rises and then cools.

Cloudiness

Overcast summer days are cooler than sunny days with clear skies because the clouds stop some of the Sun's energy reaching the ground.

Are weather forecasts always accurate?

Forecasts for the next few days are usually reliable. Predicting further ahead is trickier. That's because a small difference in conditions today may gradually increase and have an unexpected impact on the weather in the coming weeks.

QUICK QUIZ

1. Which can hold more moisture, cold air or warm air?

2. What do we call rain, snow, sleet, and hail?

3. Is an overcast summer day warmer or cooler than a clear summer day?

See pages 132–133 for the answers

Is there weather on other planets?

Any planet with an atmosphere will have weather of some kind. The gas giants (Jupiter, Saturn, Uranus, and Neptune) are mostly only atmosphere around a small, solid core. They have powerful winds and violent storms. Rocky planets (Venus, Mars, and Earth) have less turbulent weather. Mercury has no real atmosphere, so it doesn't really have weather.

NEPTUNE HAS THE **SOLAR SYSTEM'S FASTEST WINDS**, AT UP TO 2,000 KPH (1,240 MPH)!

Mercury

This planet is so close to the Sun that temperatures can reach a roasting 430°C (800°F). At night, the heat escapes and temperatures plunge to as low as -180°C (-290°F).

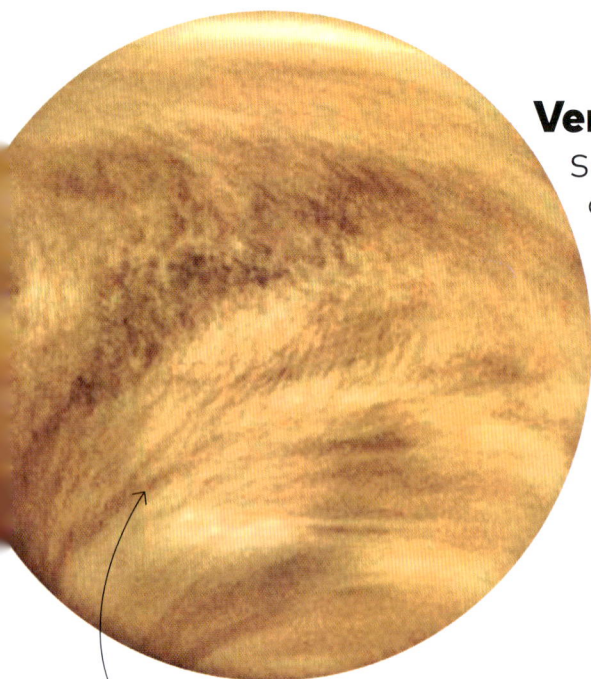

Venus

Sulphuric acid clouds and a thick atmosphere of carbon dioxide trap heat from the Sun. At the surface of Venus, temperatures are even hotter than on Mercury in the daytime.

Sulphuric acid clouds hide the surface from view.

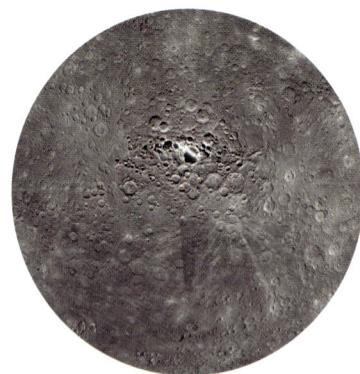

Clouds are rare on Mars.

Mars

Mars, the "Red Planet", has a thin atmosphere, but it still has winds. Martian winds stir up dust storms that can last for weeks. Occasionally, they even cover the whole planet.

Does it rain on the Sun?

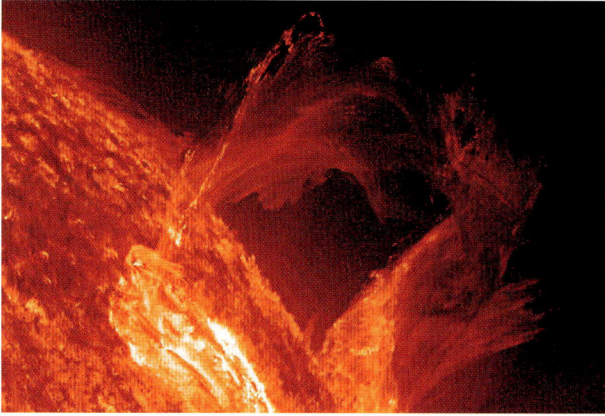

Amazingly, yes! Loops of hot, gas-like plasma rise from the Sun's surface into its upper atmosphere, which is called the corona. As the plasma cools, it falls back down to the surface. This is called coronal rain.

Saturn

A hexagonal (six-sided) whirlwind, twice as wide as Earth, spins the clouds high over Saturn's north pole. Scientists think that it rains tiny diamonds on Saturn, and on Jupiter, Uranus, and Neptune, too!

The layers of upper clouds are made of ammonia ice crystals.

Jupiter

Jupiter is wrapped in bands of swirling clouds made up of a toxic chemical called ammonia. Its Great Red Spot is a vast storm that has been raging for more than 300 years!

Great Red Spot

QUICK QUIZ

1. Which of these planets does not have weather?
   a) Jupiter
   b) Saturn
   c) Mercury

2. What are clouds made from on Jupiter?
   a) Ammonia
   b) Water
   c) Sulphuric acid

See pages 132–133 for the answers

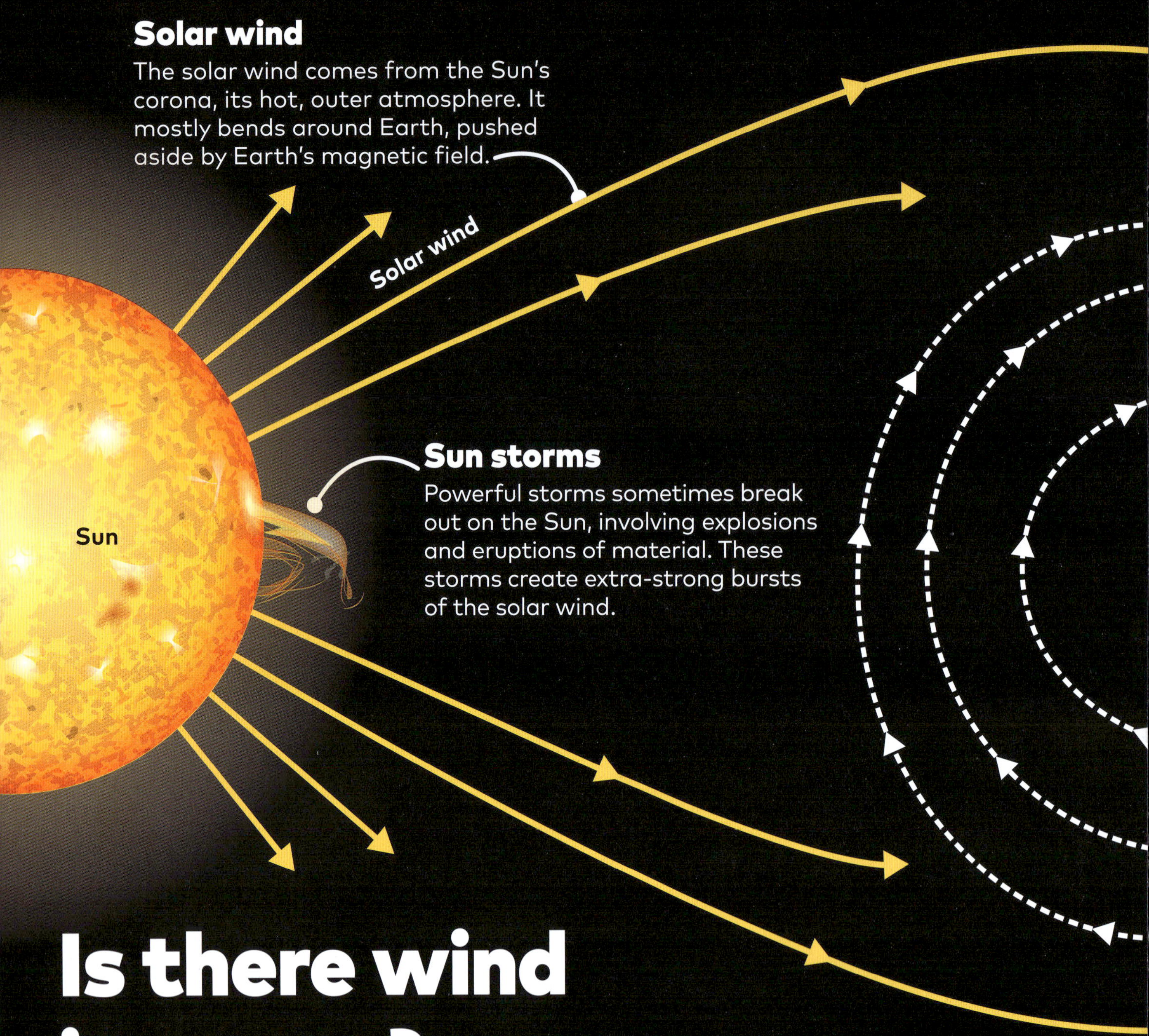

Solar wind

The solar wind comes from the Sun's corona, its hot, outer atmosphere. It mostly bends around Earth, pushed aside by Earth's magnetic field.

Solar wind

Sun

Sun storms

Powerful storms sometimes break out on the Sun, involving explosions and eruptions of material. These storms create extra-strong bursts of the solar wind.

Is there wind in space?

Yes, it's called the solar wind, but it is unlike wind on Earth. The solar wind is a stream of tiny particles, charged with electricity, that gush non-stop from the Sun into space. These particles enter Earth's atmosphere above the poles and trigger amazing light displays in the sky, called auroras.

SOLAR WIND PARTICLES HURTLE THROUGH SPACE AT ABOUT 1.4 MILLION KPH (0.9 MILLION MPH)!

Solar wind particles enter atmosphere

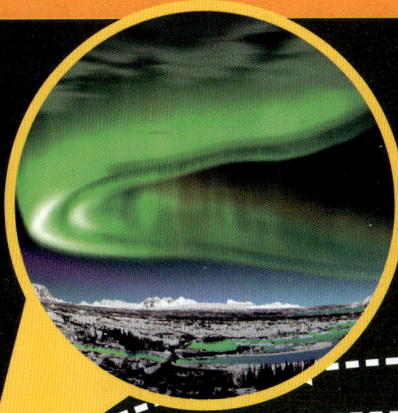

Aurora

Some solar wind particles squeeze down the Earth's magnetic field over the poles. When they slam into air molecules, the molecules give off light, creating colourful patterns in the sky.

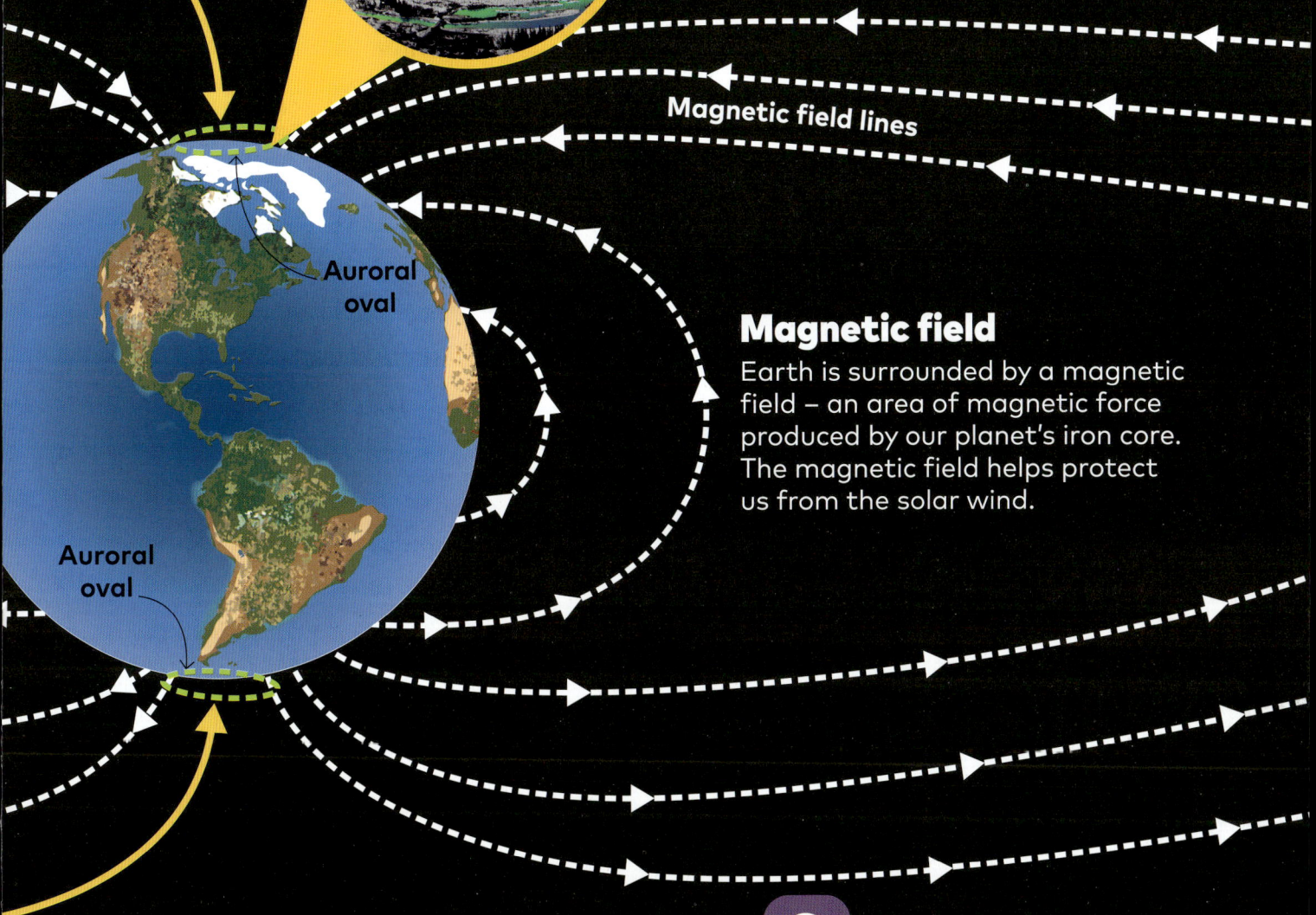

Magnetic field lines

Auroral oval

Auroral oval

Magnetic field

Earth is surrounded by a magnetic field – an area of magnetic force produced by our planet's iron core. The magnetic field helps protect us from the solar wind.

How do storms on the Sun affect us?

Solar storms send out bursts of radiation, particles, and magnetism. This "space weather" can damage satellites orbiting Earth and electricity power grids on the ground. Radiation can also be harmful to astronauts in space.

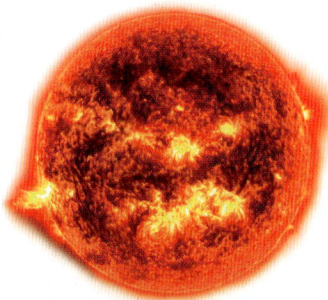

QUICK QUIZ

1. What do we call the colourful light displays caused by the solar wind?

2. From which part of the Sun does the solar wind come?

See pages 132–133 for the answers

Weather patterns

The changing weather patterns we see happen because the atmosphere is never still. The way the air moves gives us winds, from gentle breezes to powerful storms that can rip the roofs off houses. And the way water moves through the atmosphere gives us clouds, rain, snow, hail, fog, and mist.

2. Vapour condenses

The vapour condenses onto miniscule, floating particles, such as dust, pollen, or grains of sea salt. It can either form tiny water droplets that are 100 times smaller than raindrops, or ice crystals.

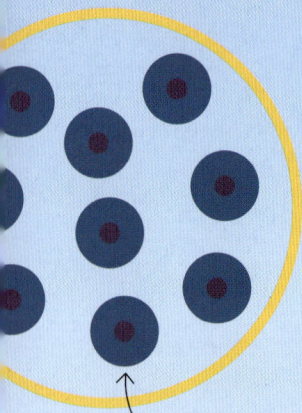

Water droplets condensed around dust particles

Water vapour

1. Air rises

Moist air warmed by the ground expands and becomes lighter, so it floats upwards. It gets cooler as it rises, causing the invisible water vapour it carries to condense.

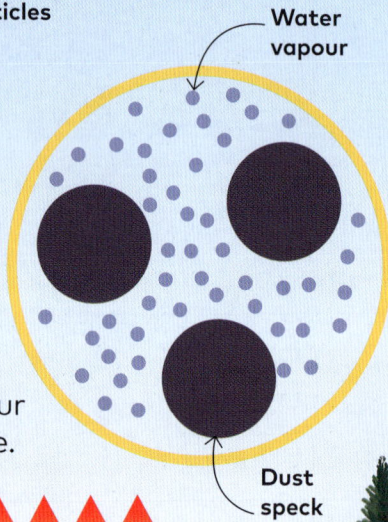

Dust speck

3. A cloud is born

When billions of water droplets or ice crystals form together, a cloud develops. This sort of fluffy, heaped cloud is called a cumulus cloud.

Cumulus cloud

How do clouds form?

When warm air rises and cools, the water vapour it holds condenses into clouds of water droplets or ice crystals. Clouds also form when warm, moving air is pushed up by hills or mountains, or is forced upwards by colliding air masses.

A CUMULUS CLOUD 1 KM (0.6 MILES) LONG, TALL, AND WIDE, COULD HOLD AROUND 500 TONNES (550 TONS) OF WATER!

4. Growing up

As more warm air rises, the cloud gets taller and wider. If a huge amount of warm air rises at once, it may grow into a towering cumulonimbus cloud.

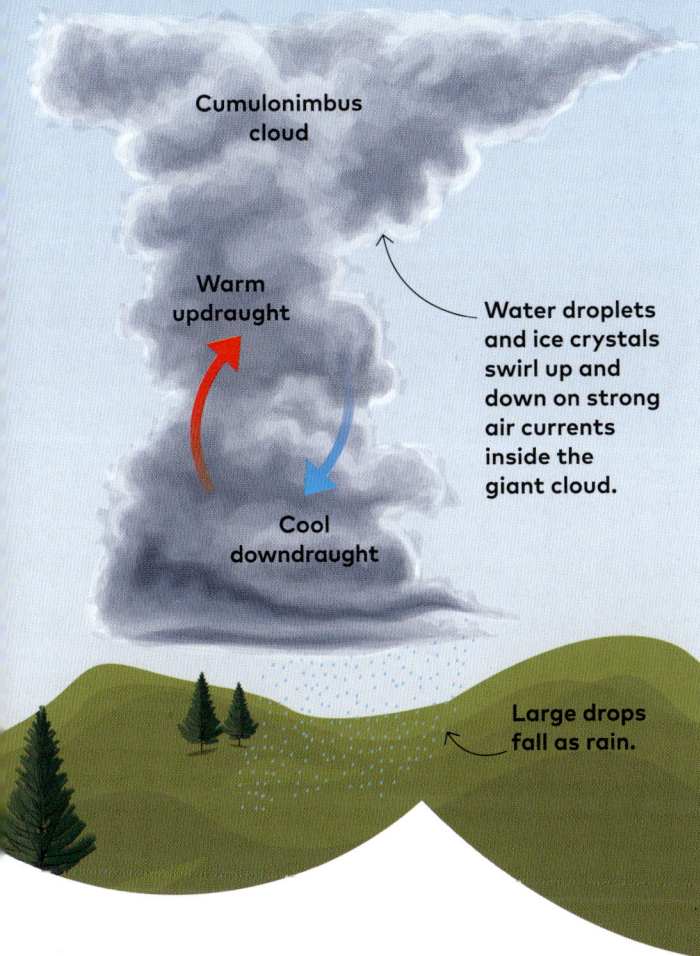

Cumulonimbus cloud

Warm updraught

Cool downdraught

Water droplets and ice crystals swirl up and down on strong air currents inside the giant cloud.

Large drops fall as rain.

5. Cloud's end

Eventually, air stops rising and creating more of the cloud. The cloud mixes with warm, dry air, and its droplets start to evaporate. It gets thinner, wispier, and gradually disappears.

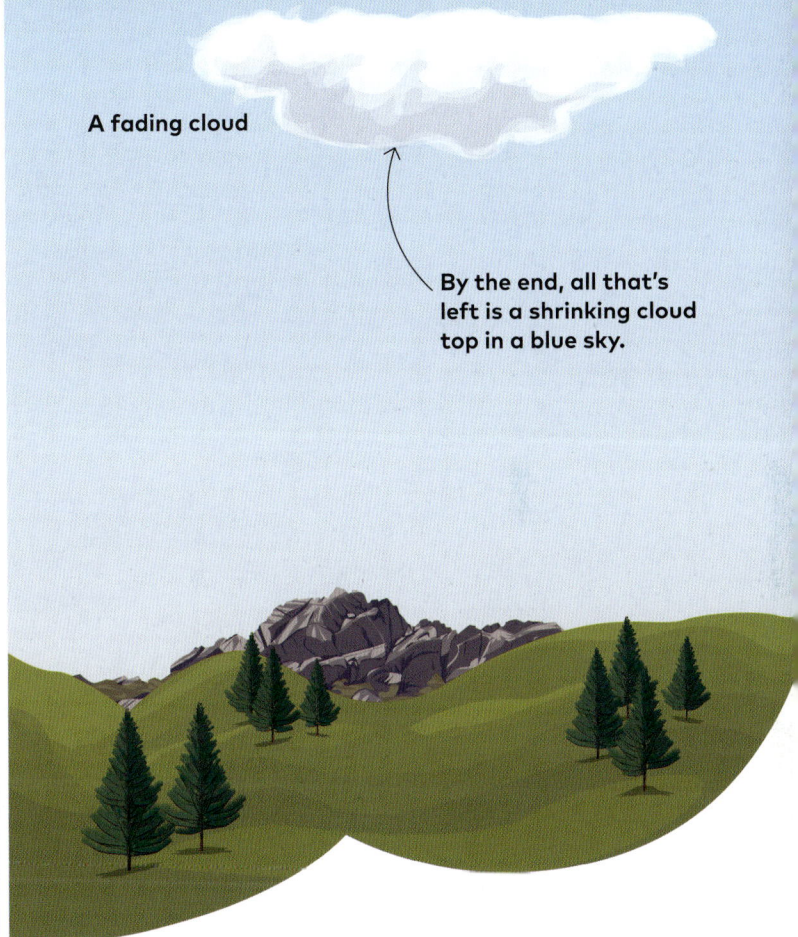

A fading cloud

By the end, all that's left is a shrinking cloud top in a blue sky.

Why do jet aircraft make clouds?

Water vapour from a jet's engine exhaust condenses and freezes, forming streaks of cloud called contrails. In dry air, contrails quickly disappear, as the ice crystals turn back to vapour. In humid air, they hang around as long, fluffy trails.

QUICK QUIZ

1. What are clouds from a jet aircraft called?

2. What does water vapour condense onto in the air?

3. What is a fluffy, heaped cloud called?

See pages 132–133 for the answers

Why are clouds different shapes?

A cloud's shape depends on the conditions in which it forms, its height in the sky, and how it's pulled and pushed by moving air. Clouds formed in smooth-flowing air form flat layers, while more restless air gives "lumpy" or "piled up" clouds. Strong winds stretch high-level clouds into wispy strands.

Cirrocumulus
Small, puffy cirrocumulus clouds make the sky look dappled or rippled. These ice-crystal clouds sometimes create a scaly pattern called a "mackerel sky".

Altocumulus
Puffs or rolls of altocumulus clouds form at medium height. Sometimes called cloudlets, these small ice-and-water clouds are seen in settled weather.

What do cloud names mean?

The Latin names of clouds describe their shape. Cirrus means a curl of hair, stratus means a layer, and cumulus a heap. Nimbus means rain cloud, and alto means high. We can put these names together to describe other clouds. Cumulonimbus, for example, means "heaped rain clouds".

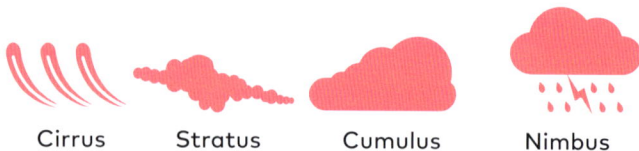

Cirrus   Stratus   Cumulus   Nimbus

Nimbostratus
Dark, grey nimbostratus clouds are thick enough to block out the Sun. They usually produce hours of rain or snow.

Stratus
Low-level, sheet-like stratus clouds appear at ground level in the form of fog. Sometimes covering the whole sky, they produce light rain or drizzle.

High level

Cirrus

Cirrus are ice clouds whipped into thin, white streamers by winds more than 6 km (4 miles) high. They are often seen in fine weather, but they may signal a coming storm.

Cirrostratus

Thin, sheet-like cirrostratus clouds are made of ice crystals. They may blanket the whole sky, and usually appear a day or so before rain or snow.

Mid level

Altostratus

Sheets of grey, mid-level altostratus clouds are made of water and ice. They can thicken, sink, and become rainy nimbostratus clouds.

Low level

Stratocumulus

Low, lumpy, and grey, stratocumulus clouds can blanket the whole sky. They occasionally bring light drizzle, but usually the weather stays dry.

Cumulus

Fluffy, white, flat-bottomed cumulus clouds form when warm air rises on a sunny day. They sometimes grow into cumulonimbus clouds.

Cumulonimbus

These huge, heaped clouds start at a low level, but can mushroom to 15 km (10 miles) tall! They can bring heavy rain, thunder and lightning, and hail.

PICTURE QUIZ

Is this a cloud or a UFO?

See pages 132–133 for the answers

Is air wet?

Air doesn't feel wet, but it does contain moisture in the form of invisible water vapour, as well as floating water droplets and ice crystals in clouds. Earth's water is always moving between the air, land, and sea in an endless loop called the water cycle.

Water is the only substance that can be found naturally on Earth as a solid, liquid, and gas.

Why does it sometimes feel "sticky"?

Humidity is a measure of how much water vapour the air is holding. The weather feels "sticky" or "muggy" when it's hot and humidity is high. The air can't hold much more water, so it is harder for sweat to evaporate from your skin.

Evaporation

The heat of the Sun causes water on Earth's surface to turn into water vapour, which is a gas. Changing from liquid to gas is called evaporation. The water vapour rises into the air, and cools.

Collection

A lot of precipitation falls at sea. On land, rain and melted snow run into rivers or underground. Some collects in lakes or is taken up by plants, but most eventually flows to the sea.

Condensation

The vapour turns into water droplets. Changing from gas to liquid is called condensation. The droplets float and form clouds. In very cold air, they may freeze to ice.

What is transpiration?

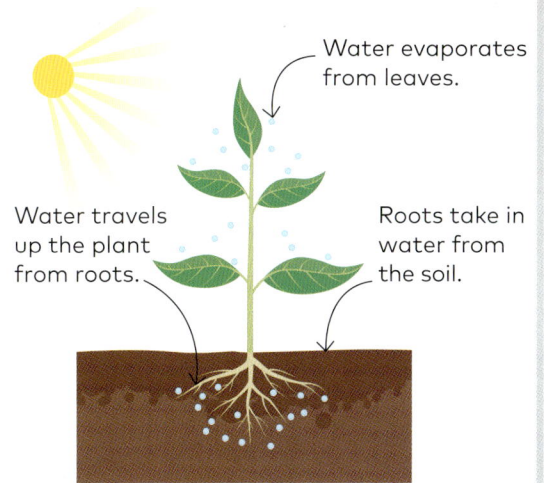

Water evaporates from leaves.

Water travels up the plant from roots.

Roots take in water from the soil.

Transpiration is the way water moves through plants. As a plant's leaves lose water to the air by evaporation, more water is drawn up through the plant from the ground. In this way, trees and other plants add moisture to the air.

PICTURE QUIZ

When water droplets form on a can of cold drink taken from the fridge, what is happening?

a) Evaporation

b) Condensation

c) Freezing

See pages 132–133 for the answers

Precipitation

The droplets or ice crystals in clouds join up, making bigger ones. If they become too heavy, they fall as drizzle, rain, sleet, snow, or hail. This is called precipitation.

Is fog the same as mist?

Yes and no! Mist and fog are simply clouds that form at ground level. Ground-level clouds are called fog when you can see less than 1,000 m (3,000 ft) in front of you. Mist is not as thick as fog, so you can see further through it.

Fog

Cold air holds less water vapour than warm air. Fog forms when warm air cools so that it can't hold its water vapour any more. The water vapour condenses into droplets of liquid water that hang in the air as fog.

What is smog?

Smog is fog mixed with air pollution. Today, air pollution is mostly from vehicle exhausts. In the 19th and 20th centuries, smogs were caused by burning coal in homes and factories. London had very thick, green-yellow smogs called pea-soupers. Here, you can see a London policeman wearing a mask for smog protection in the 1960s.

San Francisco, USA, is known for its summer fogs. They roll in from the sea when warm, moist air blows over the cool ocean.

Mist

Mist may form over low ground after a clear night. Like fog, mist clears when sunshine warms the air and turns the liquid droplets back into invisible water vapour, through the process of evaporation.

TRUE OR FALSE?

1. Ground-level clouds are called fog when you can see less than 100 m (300 ft) in front of you.

2. Smog is a mix of fog and air pollution.

3. Haze is caused by dust and water droplets in the air.

See pages 132–133 for the answers

Mist usually clears more quickly than fog, and it is easily blown away by a light breeze.

What is haze?

Haze makes far-off objects look fuzzy, especially on warm days. It is caused by dust and other tiny particles in the air, not by water droplets. Although tiny, the particles are big enough to make the air look less clear.

Is drizzle the same as rain?

Both drizzle and rain are types of precipitation made up of water droplets, but there is a difference. Drizzle droplets are very small, less than 0.5 mm (0.02 in) across. They are just heavy enough to fall from clouds. Anything larger is a raindrop.

Drizzle

Drizzle usually falls from low-level stratus clouds. It is more common in hilly areas, where clouds are closer to the ground.

What causes rain to fall?

In warm clouds, air currents hurl water droplets together. The droplets join and grow into bigger drops. They fall when they become too heavy to float. Rain also falls when snow from cold clouds melts as it falls through warmer air.

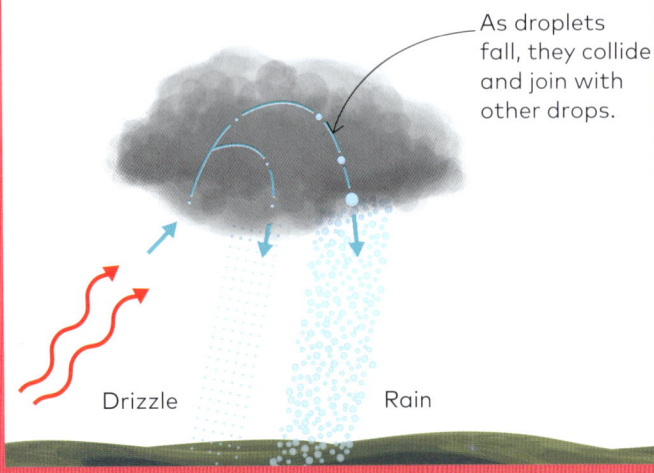

As droplets fall, they collide and join with other drops.

Drizzle   Rain

THE **LARGEST RAINDROP** EVER RECORDED MEASURED AT LEAST **8.6 MM (0.3 IN)** ACROSS.

1. If a falling drop of water is less than 0.5 mm (0.02 in) across, is it rain or drizzle?
   a) Drizzle
   b) Rain

2. Which of these is the correct order for the changing shape of raindrops?
   a) Bun, kidney, ball, bell, ball
   b) Ball, bell, bun, kidney, ball
   c) Ball, bun, kidney, bell, ball

See pages 132–133 for the answers

What shape are raindrops?

Small drops are ball-shaped. As they grow bigger and fall, the air pushes up on them and their bottoms flatten out. If a drop reaches about 5 mm (0.2 in) wide, it usually stretches into a bell shape then splits into two ball-shaped drops.

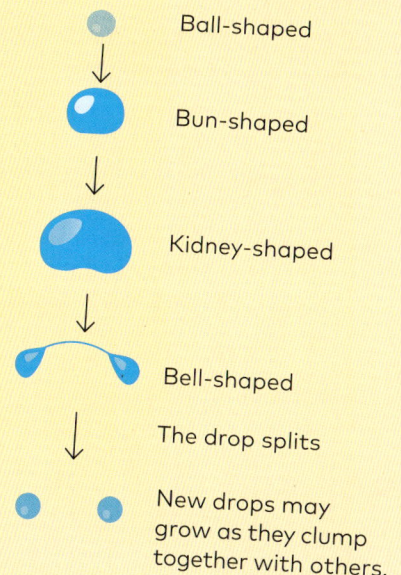

Ball-shaped

Bun-shaped

Kidney-shaped

Bell-shaped

The drop splits

New drops may grow as they clump together with others.

Rain

A typical raindrop is about 2 mm (0.08 in) across – four times bigger than a drizzle droplet. Being larger and heavier than drizzle droplets, raindrops fall faster.

What is a rainbow?

A rainbow is an arc of colours seen in the sky when the Sun shines on a rainy day. Although sunlight looks white, it is really a mixture of different colours. As sunlight passes through raindrops, it splits out into these different colours, which fan out into a rainbow.

GHOSTLY "FOGBOWS" CAN FORM WHEN SUNLIGHT HITS TINY WATER DROPLETS IN FOG.

Seeing a rainbow

To see a rainbow, you must have the Sun behind you, the rain in front of you, and the Sun must not be too high. The lower the Sun is, the wider the rainbow.

Colour parade

The colours in a normal rainbow always appear in the same order. Red is at the top, then, orange, yellow, green, blue, and indigo, with violet at the bottom.

Why do we see rainbows?

Sunlight entering a raindrop reflects off the inside surface of the drop back to your eyes. The different colours in sunlight bend by different amounts as they enter and then leave the drop. This makes them separate into a spectrum.

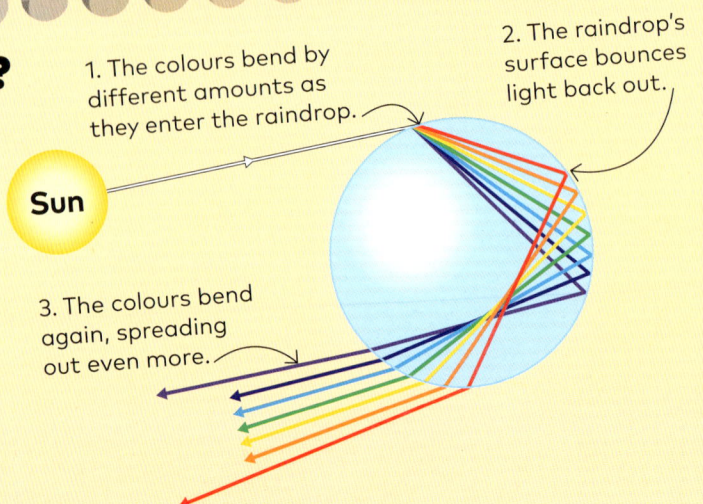

1. The colours bend by different amounts as they enter the raindrop.

2. The raindrop's surface bounces light back out.

Sun

3. The colours bend again, spreading out even more.

Do rainbows form around the Sun?

No, but rings of light called haloes can form around the Sun and Moon. They happen when sunlight or moonlight reflects off ice crystals in high cirrus clouds. The haloes are usually white, but they can be faintly coloured, too.

QUICK QUIZ

1. How should the Sun be positioned, in order to see a rainbow?
   a) In front of you
   b) High in the sky
   c) Behind you

2. Which of these is the correct order of rainbow colours?
   a) Red, orange, yellow, green, blue, indigo, violet
   b) Blue, indigo, violet, red, orange, yellow, green
   c) Violet, indigo, blue, yellow, green, red, orange

See pages 132–133 for the answers

Double rainbow

If sunlight reflects twice inside raindrops, you see a second, fainter rainbow higher in the sky. The colours in the second rainbow are in reverse order, starting with violet at the top.

Rainbow's end

A rainbow looks like it ends on the horizon, but it would be a full circle if the ground wasn't in the way – we can only see part of it. High-flying pilots sometimes see a full-circle rainbow!

What causes lightning?

Inside a thundercloud, ice crystals and freezing water droplets crash into each other and become electrically charged. Lower parts of the cloud become negatively charged while upper areas gain a positive charge. These opposite charges create a big spark of electricity in the cloud, or between the cloud and ground, that we see as lightning.

LIGHTNING CAN HEAT AIR TO TEMPERATURES UP TO 30,000°C (54,000°F)!

Zigzagging down

A positive charge builds up on the ground and attracts the negative charge at the cloud's base. A sudden flow of charge zigzags from the cloud to the ground faster than the eye can see, creating an electrically charged pathway in the air.

Flashing back

A split second later, a massive surge of lightning shoots back up the jagged pathway to the cloud. The lightning bolt lights up the sky and heats the air around it.

What happens when lightning strikes a tree?

When lightning strikes a tree, the massive amount of heat turns the watery liquid, called sap, into gas. This gas expands and splits the tree from inside. Usually, only the bark is affected and the tree survives. But sometimes, the tree gets blown apart!

Electrically charged clouds

Most lightning actually happens inside of clouds. Positively charged ice and water are lighter, so they tend to rise towards the top of the cloud. The negative charges are heavier, and gather at the base.

Thunder

Lightning heats and expands air, producing a shock wave, called thunder. Because sound travels slower than light, we hear the sound of thunder only after we see the lightning flashing.

QUICK QUIZ

1. How hot can air get when lightning strikes?

2. Where does lightning mostly happen?

3. What two things are hurled together inside a thundercloud?

See pages 132–133 for the answers

Why is the sky blue?

The sky looks blue because gas molecules and particles in the atmosphere scatter sunlight. Light travels as waves, and each of the colours that make up sunlight has waves of a different length. The blue end of the spectrum has shorter waves, and is scattered most, making the sky appear blue.

Why not violet?

Violet and indigo are scattered more than blue, but this happens higher in the atmosphere, so we don't see them. Our eyes are also more sensitive to blue light than to violet or indigo.

All blue

Blue light is scattered in all directions around the sky. Whichever way you look, you will see blue light.

Air particles scatter light

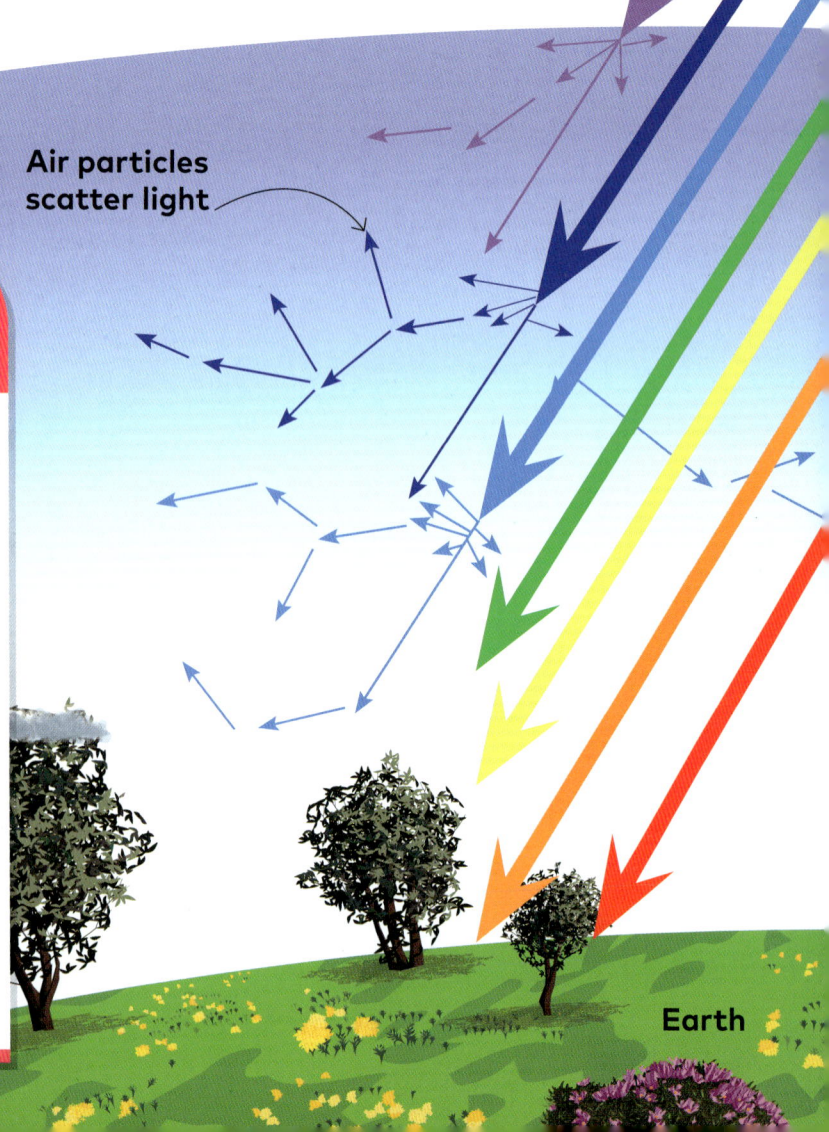

Why is the horizon less blue?

The sky looks less blue on the horizon than overhead. That's because some scattered blue light coming from the horizon gets scattered again. The extra scattering means less blue light reaches your eyes, so the horizon is paler.

Earth

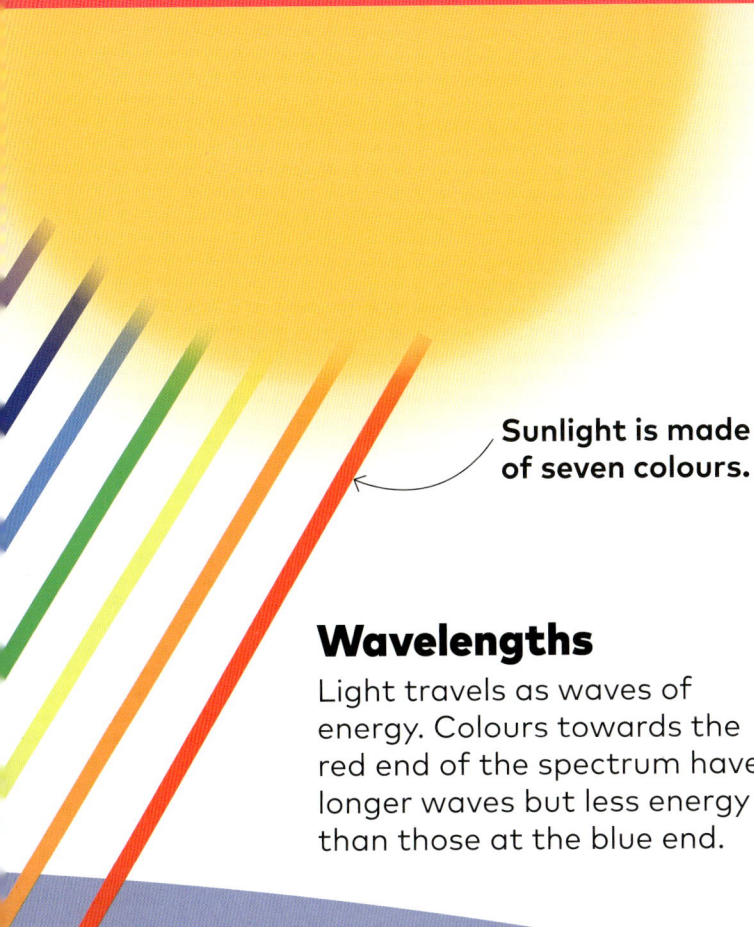

Sunlight is made of seven colours.

Wavelengths
Light travels as waves of energy. Colours towards the red end of the spectrum have longer waves but less energy than those at the blue end.

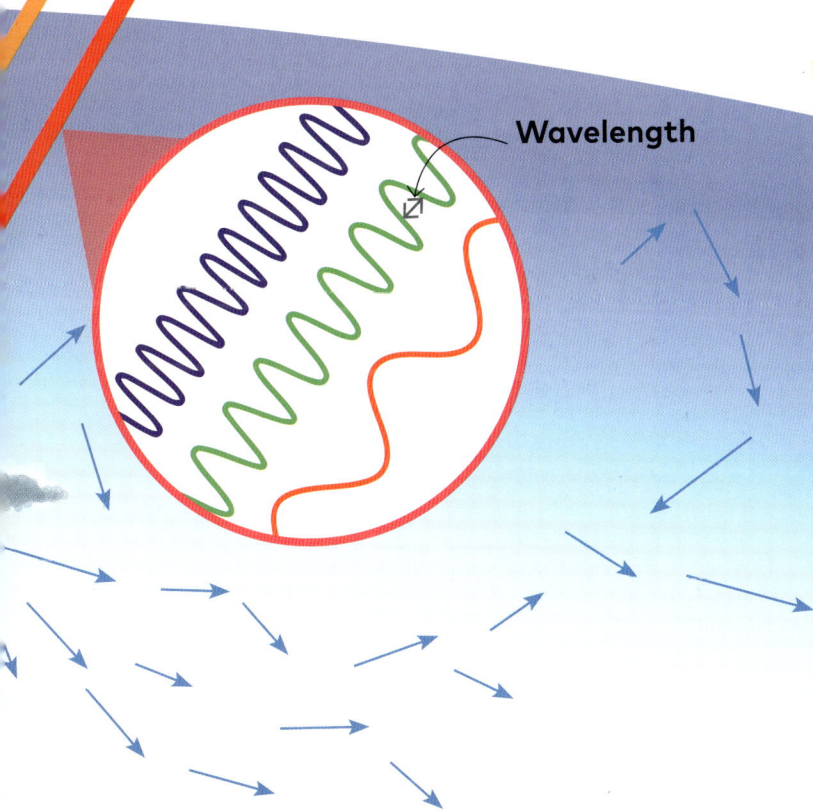

QUICK QUIZ

1. Which colour light is scattered in all directions around the sky?
   a) Yellow
   b) Green
   c) Blue

2. Which colour light scatters the most?
   a) Violet
   b) Red
   c) Green

See pages 132–133 for the answers

Wavelength

Why do we get colourful sunsets?
When the Sun hangs low in the sky, sunlight must pass through much more atmosphere before it gets to you. Blue and other shorter waves are completely scattered away. Only longer waves – yellow, orange, and red – get through, giving dazzling sunsets.

Atmosphere

Why does the wind blow?

Winds blow because of the uneven way the Sun heats up Earth's surface. For example, the land heats up faster than the ocean. Warm, low-pressure air over land rises, and cooler, higher-pressure air rushes off the ocean to take its place, creating a wind called a sea breeze.

What are winds like in hilly regions?

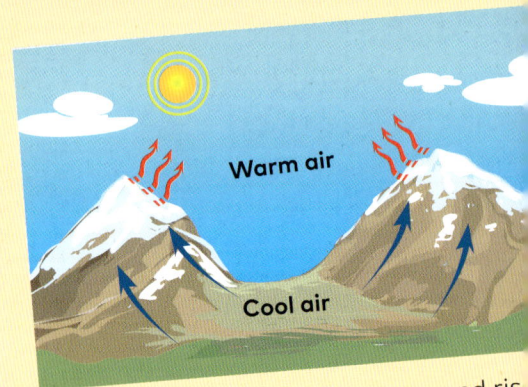

By day, air over the slopes warms and ris Cool air from the valley below flows up th slopes as a valley wind.

What is a land breeze?

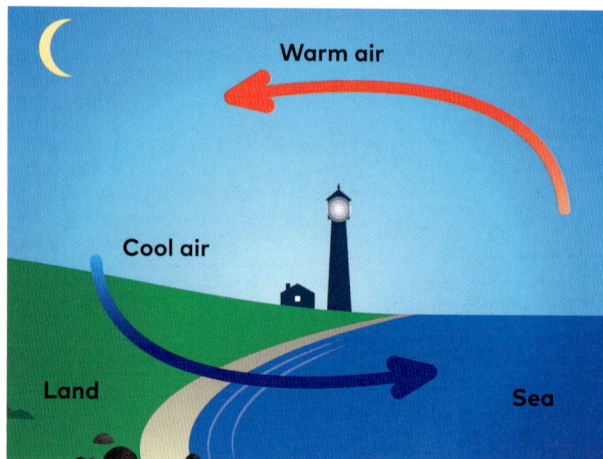

A night-time wind blowing from the land out to sea is called a land breeze. It happens in the opposite way to a sea breeze. After the Sun sets, the land cools faster than the ocean. Cool air over the land sinks, and is drawn seawards.

1. Rising air

The air over land is warmed by the ground. It expands, becomes lighter, and rises. A low-pressure area, where the air is less squished together, forms over the land.

Cool air sinks as it becomes heavier.

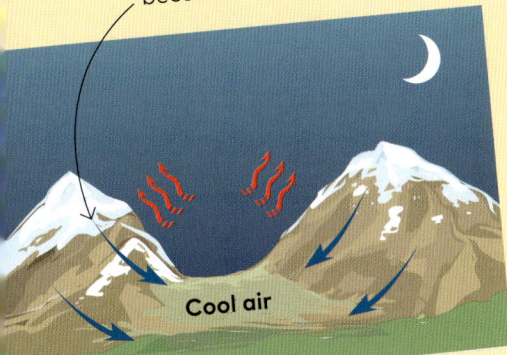

Cool air

When air cools above mountain slopes or hillsides at night, it drains downhill to the valley as a mountain wind.

1. Sea breeze winds blow from high-pressure to low-pressure areas.

2. A land breeze is the night-time opposite of a sea breeze.

3. During the day, cool air from hill and mountain slopes flows towards the valley.

See pages 132–133 for the answers

2. Offshore wind

The warm air cools as it rises. It starts to become denser (more squished together), so its pressure increases as it flows towards the ocean.

3. Sinking air

Once over the ocean, the air keeps cooling. As it cools, the air becomes even more dense and heavy, so it sinks.

4. Onshore wind

Cool, dense, higher-pressure air is pushed towards the land to take the place of the warm air that rises. It flows as a wind called a sea breeze.

Winds flow from high-pressure areas to low-pressure areas.

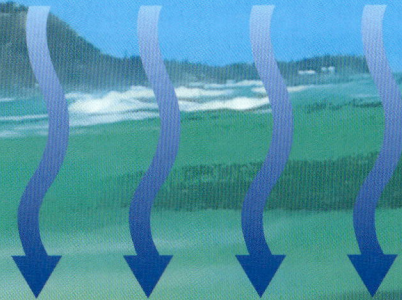

How strong does the wind blow?

The strength, or force, of the wind is how fast the air is moving. The strongest winds, which come with hurricane storms, blow at more than 118 kph (73 mph). The Beaufort scale typically divides wind strength into 13 named levels, based on what you can see happening.

1. Calm
Smoke rises straight upwards. The sea looks smooth, like a mirror.

Who invented the Beaufort scale?

The Beaufort scale was devised in 1805 by Francis Beaufort to help sailors judge wind strengths at sea. It was later applied to winds on land, too. Next time you're out, see if you can use it to guess the strength of the wind.

5. Moderate breeze
Dust and loose paper lift off the ground, and small branches move. Small waves develop at sea, some with white crests.

QUICK QUIZ

1. Why was the Beaufort scale invented?

2. How many levels of wind strength are there according to the Beaufort scale?

See pages 132–133 for the answers

9. Gale
Twigs snap off trees, and it is almost impossible to walk into the wind. Waves may reach 7.5 m (25 ft) in height. Their crests turn to spray.

10. Strong gale
Branches break. Roof slates and tiles are blown off houses. Out at sea, wave crests topple and roll over.

2. Light air

Smoke drifts gently, showing the wind's direction. Ripples form on the surface of the sea.

3. Light breeze

Leaves rustle and you can feel the wind on your face. Wavelets (mini-waves) up to 30 cm (1 ft) high appear at sea.

4. Gentle breeze

Leaves and small twigs move in the breeze, and light flags flutter. Larger wavelets form on the ocean.

6. Fresh breeze

Small trees start to sway. Ocean waves are medium-sized, up to 2.4 m (8 ft) high, and many have white crests.

7. Strong breeze

Large branches move, and it is difficult to keep an umbrella open. Ocean waves are large, with big, foamy crests.

8. Near gale

Whole trees sway, and it's hard to walk into the wind. Streaks of foam from breaking waves blow over the sea.

11. Storm

Houses are damaged and trees are uprooted. The sea looks white. Huge waves up to 12.5 m (41 ft) tall have overhanging crests.

12. Violent storm

Serious damage is done to buildings. At sea, visibility is poor, and exceptionally high waves may hide smaller ships from view.

13. Hurricane

Towns and cities on the coast may be devastated. There are mega-waves over 13.7 m (45 ft) tall at sea, and the air is filled with foam and spray.

1. Ice crystal

Every snowflake begins with water vapour freezing onto a pollen grain or dust particle in a cloud. It forms an ice crystal shaped like a hexagonal (six-sided) prism.

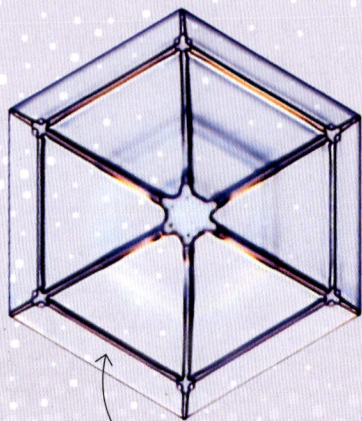

2. Growing arms

More water vapour freezes onto the ice crystal as it swirls inside the cloud. Arms begin to grow from the corners as more water is added.

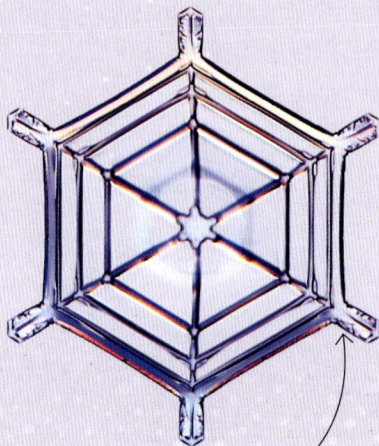

3. Branching out

The snowflake's arms sprout side branches. A snowflake is always symmetrical – if you could cut it in half, each half would look identical.

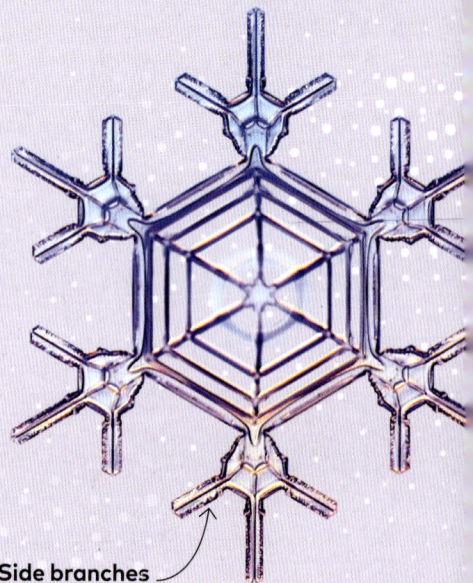

A six-sided ice crystal forms

Arms start to grow from corners

Side branches start forming on arms

Is every snowflake different?

No two snowflakes have the same shape. A snowflake's shape depends on the temperature and amount of moisture in the air when it forms. Each snowflake follows a different path through the air, meeting slightly different conditions on the way, and these conditions give it its own special shape.

4. Different shapes

If the snowflake forms quickly, the arms become tree-like. If it forms slowly, there may be flat plates in the arms. Snowflakes collide, clump together, and fall to the ground when they get too heavy.

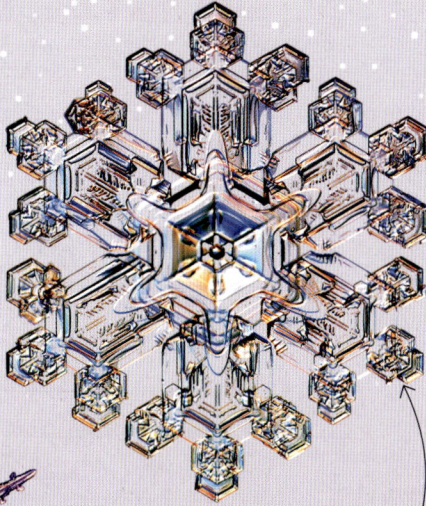

All snowflakes have six arms because of the way water molecules link together in ice.

See pages 132–133 for the answers

QUICK QUIZ

1. How many arms does a snowflake have?

2. Why does snow look white?

3. What type of snow forms in cooler, drier air?

Why is some snow fluffy and some powdery?

Snowflakes falling through warmer, moister air stick together to make large, fluffy flakes that are great for making snowmen. In colder, drier air, snowflakes don't stick together so well. This makes powdery snow that's ideal for winter sports.

What colour is snow?

Falling snowflakes and snowflakes on the ground look white, but they are really colourless. Snowflakes reflect the sunlight that falls on them back to our eyes, and because sunlight looks white, so does the snow.

Ground frost

If the cold ground chills the air above it to below 0°C (32°F), the ground soon becomes covered with a layer of white ice crystals that look like sugar granules.

Hoar frost

When cool air blows over ice-cold objects, its moisture freezes and coats branches, leaves, and metal surfaces with glittering ice needles of hoar frost.

Are frost and ice the same?

All frost is ice, but all ice isn't frost! Frost is ice crystals that grow on objects during chilly nights. The air's water vapour turns directly to ice crystals without first condensing. When cold nights freeze airborne water droplets, they coat surfaces with a layer of ice, rather than frost crystals.

FROSTBITE IS DAMAGE TO THE SKIN FROM **EXTREME COLD.** EARLY POLAR EXPLORERS OFTEN SUFFERED FROM FROSTBITE.

Fern frost

Delicate, fern-like patterns or feathery trails of ice crystals may form on windows and car windscreens as water vapour in the air freezes onto cold glass.

Rime ice

A thick, shiny coating of white ice called rime forms when extremely cold water droplets in fog are blown onto cold surfaces, and instantly freeze.

What is black ice?

Sometimes, cold rain or drizzle freezes on impact, coating things with clear ice called glaze. Glaze on roads, known as "black ice", makes driving dangerous. Black ice is either difficult to see, or makes roads look wet rather than icy. It is common in shady places, and also on bridges, where cold air flowing underneath lowers temperatures above.

PICTURE QUIZ

Are icicles frost or ice?

See pages 132–133 for the answers

SINKING AIR FLOWS CLOCKWISE IN THE NORTHERN HEMISPHERE AND ANTICLOCKWISE IN THE SOUTHERN.

Clear skies

Clouds form when warm, moist air rises. But when cold air sinks, it stops the air below from rising. No clouds can form, so the sky is clear, blue, and sunny.

Sinking cold air

Circling air

Due to Earth's spin, the sinking air circles around the centre of the high-pressure area.

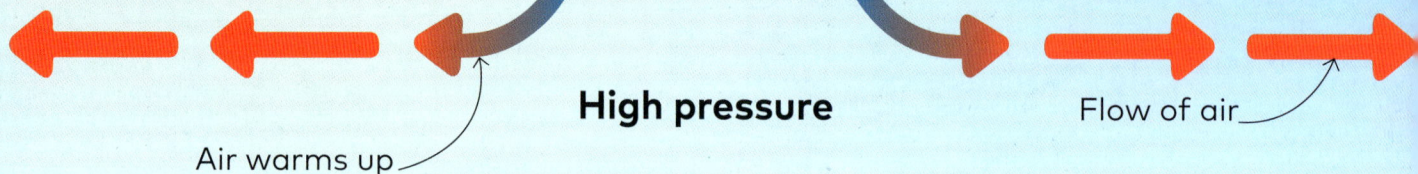

Sinking air

As the cold air sinks, its pressure rises and it gets warmer. It flows away from the high-pressure area as gentle winds.

High pressure

Air warms up

Flow of air

What makes some days sunny?

Sunny days are caused by high pressure. When the air above is relatively cold, it slowly sinks and squashes the air below, raising its pressure. The sinking air warms as it descends, so condensation can't form and we get clear skies. Low pressure brings the opposite – cloudy skies.

Cloudy skies

When warm, moist air rises, it creates an area of low pressure. The rising air cools and its moisture condenses to form clouds, giving overcast or rainy weather.

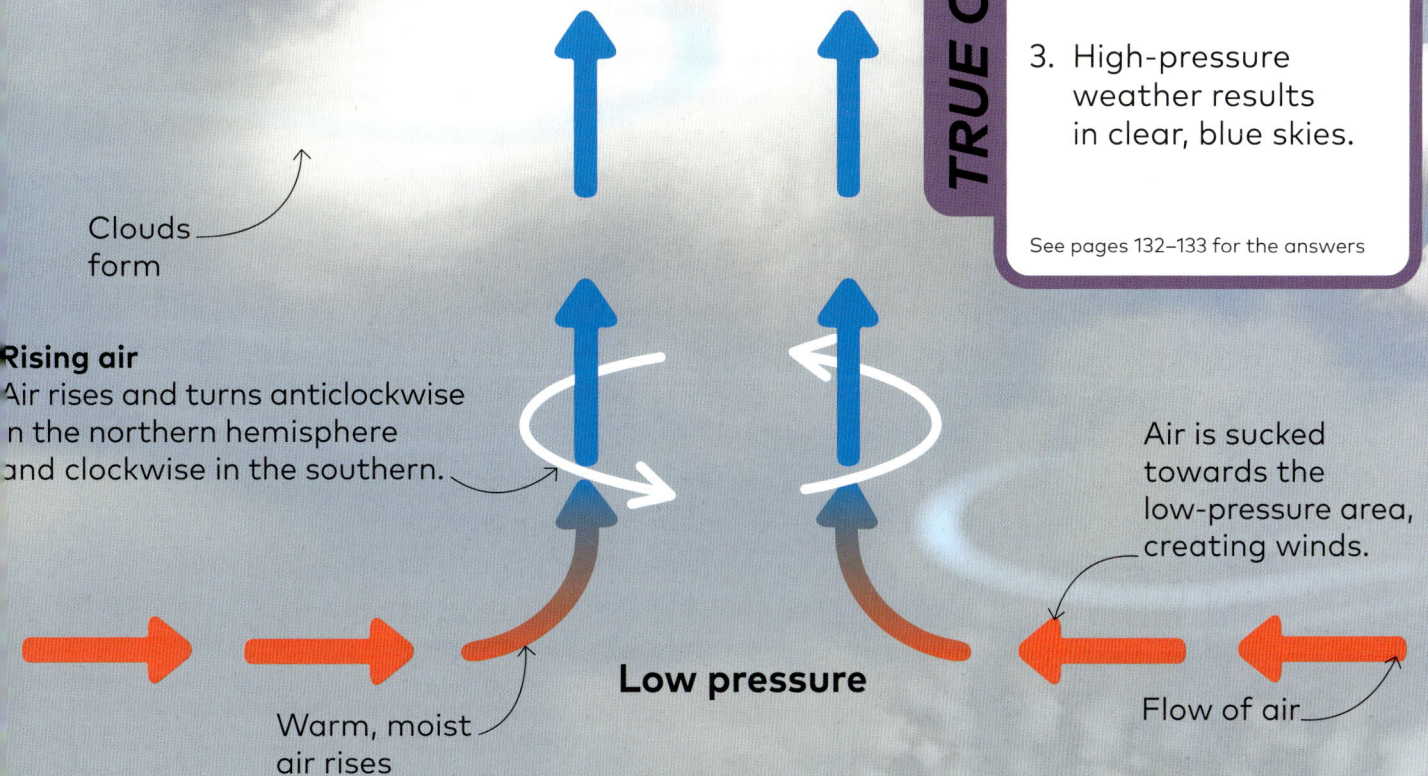

Clouds form

Rising air

Air rises and turns anticlockwise in the northern hemisphere and clockwise in the southern.

Air is sucked towards the low-pressure area, creating winds.

Low pressure

Warm, moist air rises

Flow of air

TRUE OR FALSE?

1. Clouds form when there is high pressure.

2. High pressure always brings warm weather.

3. High-pressure weather results in clear, blue skies.

See pages 132–133 for the answers

Does high pressure always mean warm weather?

In summer, high pressure often brings hot days, because there are no clouds to stop the Sun's heat from reaching the ground.

In winter, high pressure still brings clear, sunny days, but they are usually cold. Clear skies on winter nights often bring heavy frosts.

What are weather fronts?

The most unsettled weather happens at places called fronts, where two huge masses of air meet. Fronts bring wet and windy weather. They usually come as a pair – a warm front followed by a cold front – that moves along with an area of low pressure. The warm front comes first, followed by a gap, then the cold front arrives.

What is an air mass?

An air mass is a vast chunk of air that is equally cold, warm, moist, or dry throughout. Cold air masses form nearer the poles, warm ones towards the Equator. Air masses that form over oceans are moist. Dry air masses form over land. Winds move air masses around.

Warm front

A warm air mass collides with cold air and slips over it. Moisture in the warm air cools as it rises, forming thin clouds and usually bringing steady rain. As the front passes, the rain ends and it feels warmer.

COLD FRONTS MOVE MORE QUICKLY THAN WARM FRONTS, SO THEY EVENTUALLY CATCH THEM UP.

Cold, dry air mass

Warm air rises smoothly, forming sheet-like clouds.

Warm, moist air mass

What do weather maps show?

On weather maps, cold fronts are shown by a line with blue spikes, and warm fronts by a line with red bumps on it. The spikes and bumps show the direction the fronts are moving in. Lines called isobars link places with the same air pressure. Isobars form rings around areas of high and low pressure.

L marks the centre of the low-pressure area.

Fronts often revolve anticlockwise around areas of low pressure.

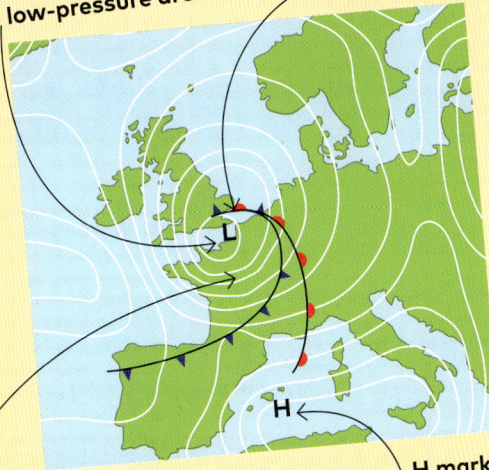

It will be windiest where the isobar lines are closest together.

H marks the centre of the high-pressure area.

Cold front

A cold air mass collides with warm air. As the warm air is forced up, its moisture may form huge clouds, bringing heavy showers and often thunder. As the front passes, it feels colder. Most of the clouds blow away.

QUICK QUIZ

1. Which type of front brings stormy weather, warm or cold?

2. What are pressure lines on a weather map called?

See pages 132–133 for the answers

Cold, dry air mass

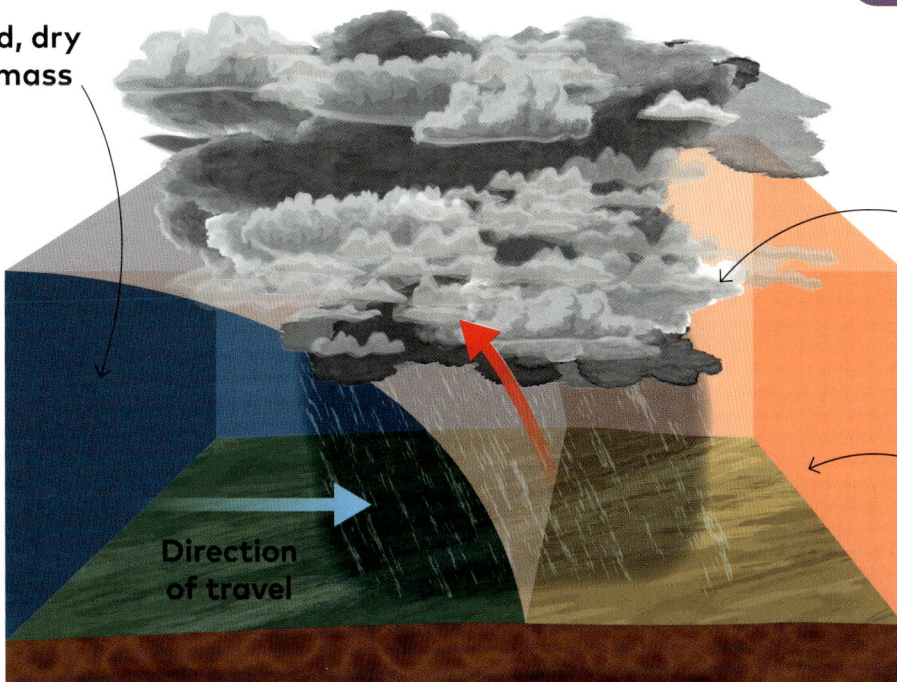

Warm air rises sharply, often forming towering thunderclouds.

Direction of travel

Warm, moist air mass

What are the strangest weather events?

Weather can be wild or calm, and sometimes just plain weird. It can create odd-shaped clouds, play tricks on our eyes, dazzle us with electrical displays, and even rain small animals from the sky. Some very strange things happen in the world of weather!

ROLL CLOUDS CAN BE UP TO 1,000 KM (620 MILES) LONG.

Ball lightning

In this rare and unexplained type of lightning, a glowing white or coloured sphere is seen floating near the ground. The sphere may move about, but soon disappears quietly or explodes harmlessly.

Frost flowers

Flower-like clusters of ice crystals grow on thin sea or lake ice. They form when ice vaporizes from the surface into cold, dry air above, and then refreezes into frosty blooms.

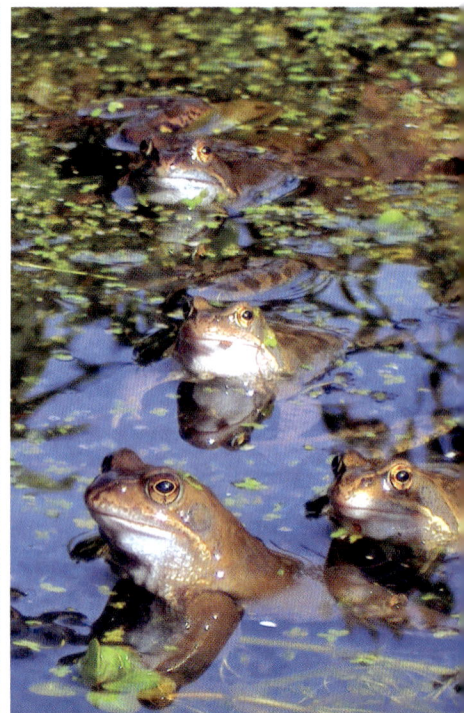

Animal rain

Occasionally, people report fish, frogs, or other small animals falling from the sky. Some scientists think the animals may get picked up by storms and dropped somewhere else.

Can mushrooms make it rain?

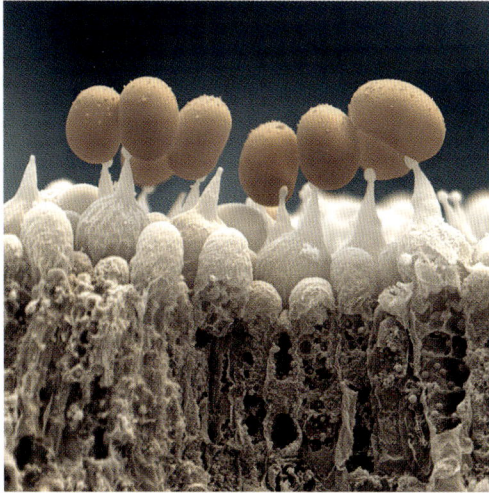

Bizarrely, yes! Each year, the wind wafts countless billions of tiny mushroom spores high into the sky. The spores help make clouds and rain because water vapour can condense around them, just as it does around floating dust grains. Airborne bacteria, algae, and pollen also make rain in this way.

Sundogs

If you think there are three Suns, you're seeing sundogs! Two bright spots may appear on either side of the Sun when sunlight passes through ice crystals in high cirrus clouds.

Roll clouds

Sometimes, long, tube-shaped clouds form in front of thunderstorms or over the sea near land. The clouds roll because the winds above blow in the opposite direction to those below.

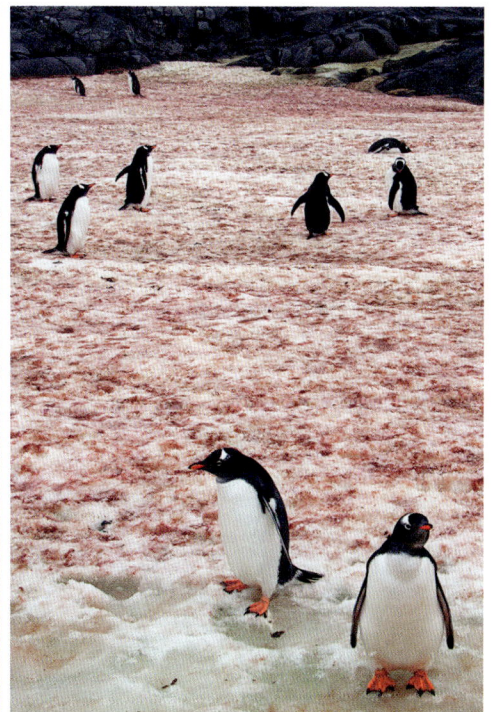

Coloured snow

Snow in mountain and polar areas may be tinted red, green, orange, and other colours by algae living in the snow. Sand, dust, and pollution can also give colour to snow.

Weather around the world

The typical weather of a place varies greatly around the world. Some places always have snow on the ground, others never see a snowflake fall. One part of the world may get lots of rain, another may be bone dry. And temperatures range from scorching hot to freezing cold.

Temperate

Mild or temperate climates have warm summers and cool winters. There can be rain at any time of the year. Temperate places nearer to the poles may have winter snow.

Temperate places have four distinct seasons and often very changeable weather.

Temperatures in tropical climates do not vary much throughout the year.

Tropical

Near the Equator, it is always hot, and there is no summer or winter. There is a lot of rainfall making tropical rainforests hot and wet.

What are climate zones?

Climate zones are large areas of Earth's surface with distinct weather patterns. The warmest climates, near the Equator, get the most sunlight. The coldest, near the poles, get the least.

THE CLIMATE OF A PLACE DETERMINES WHAT KINDS OF PLANTS AND ANIMALS CAN SURVIVE THERE.

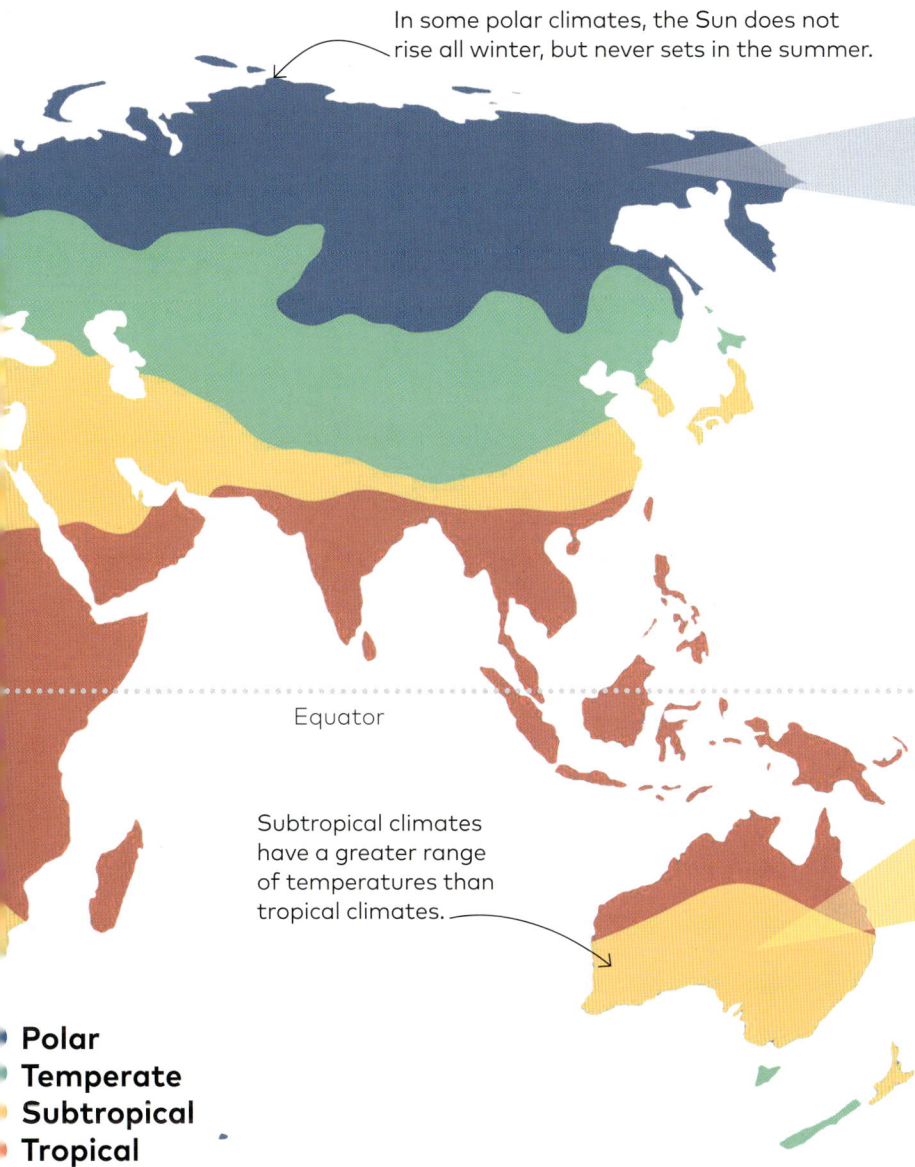

In some polar climates, the Sun does not rise all winter, but never sets in the summer.

Polar

The weather is very cold, and the snow only melts in the summer. In some polar areas, the ground beneath the soil is always frozen. This is called permafrost.

Equator

Subtropical climates have a greater range of temperatures than tropical climates.

- Polar
- Temperate
- Subtropical
- Tropical

Subtropical

Summers are often hot and winters mild. Subtropical climates may be dry or humid (damp). Much of Earth's desert lies in subtropical climate zones.

How can we compare climates?

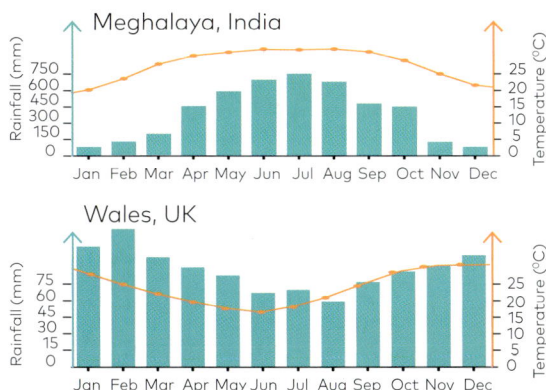

Meghalaya, India

Wales, UK

We can use graphs to compare the climates of different places. We make a graph for each, and plot the average temperatures and levels of rainfall. When we compare the graphs, it is easy to see differences.

Countries such as India, which lie in the tropical zone, see higher rainfall in the monsoon season compared to countries such as Wales, which lie in the temperate zone.

QUICK QUIZ

1. What is permafrost?

2. In which climate zones do most deserts lie?

3. How many seasons does a temperate climate have?

See pages 132–133 for the answers

What is a microclimate?

A microclimate is the climate of a small area, such as a wood, valley, or even a garden. The typical weather conditions in a microclimate are different to those of the surrounding region. An oasis, for example, has a microclimate that is different to the climate of the desert it lies in.

Cool oasis

The oasis is cooler than the desert. Evaporation from the lake and transpiration by plants take away heat and add moisture to the air. Trees provide shade.

What is an urban heat island?

An urban heat island is a built-up area that is warmer than the countryside around it. The bricks, concrete, and asphalt of cities and towns readily absorb and hold on to heat, while buildings help block cooling winds.

Huacachina, a desert oasis in Peru

Dry desert

Huge sand dunes surround the Huacachina oasis in Peru, which lies on the edge of the Atacama Desert – the driest desert in the world.

Black-crowned night heron

Wildlife haven

Microclimates are vital for wildlife. The plants growing at an oasis provide a habitat for creatures that couldn't survive in the desert dunes.

Water for life

The lake is formed by water that seeps up through the sand from underground rocks. The water enables plants to grow around the rim of the lake.

What creates a microclimate?

Slight variations in local conditions can create a microclimate. They include differences in soil, landscape, plant cover, and the amount of sunlight compared to the surrounding area. Sometimes, climate varies a lot over just a few metres. This sheltered hollow in hardened lava from a volcano has its own microclimate, which has enabled a shrub to grow.

QUICK QUIZ

1. Where does the water in Huacachina's lake come from?
   a) Rain that collects between the dunes
   b) Underground rocks
   c) A river that flows into the lake

2. What do we call a city that is warmer than the nearby countryside?
   a) A city hot-spot
   b) An urban oasis
   c) An urban heat island

See pages 132–133 for the answers

What is a drought?

A drought is a long period when little or no rain falls. Drought happens when the normal weather pattern is disrupted, and the usual rains don't come. When drought strikes, the soil dries out, crops fail, and water becomes scarce.

ANYWHERE THAT GETS **LESS THAN** 25 CM (10 IN) OF RAINFALL A YEAR IS A DESERT.

The Sahara Desert in Africa is the largest hot desert on Earth.

Hot deserts

Deserts cover about 20 per cent of Earth's land surface. In hot deserts, the Sun's energy heats the dry air, giving scorching days. At sunset, the air cools rapidly, and nights can be very cold.

Why are some places so dry?

Deserts are dry because they rarely get rain. Many deserts have high-pressure air overhead, which stops clouds forming. Others lie in the rain shadow of mountains, or are so far inland that air from the coast drops all of its moisture before it reaches them.

Cold deserts

Not all deserts are hot. Parts of the Arctic and almost all of Antarctica are desert. The cold polar air there can't hold much moisture, so there is very little precipitation, most of which is snow.

See pages 132–133 for the answers

QUICK QUIZ

1. Are hot deserts hot all the time?

2. Which is drier, the Sahara or the Atacama?

3. Why is Antarctica a dry place?

There is lots of water in Antarctica, but it is frozen in ice sheets and glaciers.

Which is the driest desert?

Some weather stations in the Atacama Desert of Chile, South America, have never recorded a drop of rain. The Atacama is the driest hot desert, but parts of Antarctica are equally dry – perhaps even drier.

What are monsoons?

Monsoons are tropical winds that change direction at the start of certain seasons. For half of the year, the dry winter, they blow out to sea. For the other half, the wet summer, they blow inland. Monsoon is also the name of the torrential summer rains that the winds bring.

Seasonal change

Monsoons happen because the land heats up and cools down faster than the sea. At the start of each monsoon season, this temperature difference causes the wind to switch direction.

Heavy rain

Summer monsoons are like massive sea breezes that blow in off the sea as the land heats up. Laden with moisture, they bring clouds and heavy downpours.

How do monsoons affect people?

Farmers in monsoon regions rely on the wet summer months for the water they need to grow crops. After the long dry season, the monsoon rains are vital for a good harvest.

When the monsoon downpours are too heavy, they can cause terrible floods. On the other hand, if the rains don't come when the farmers need them, or if they don't bring enough water, crops may fail.

QUICK QUIZ

1. When do monsoon winds bring heavy rain, in winter or summer?

2. Which is cooler in summer, the sea or the land?

3. In which direction do winter monsoon winds blow?

See pages 132–133 for the answers

How do monsoon winds form?

Monsoon winds always blow from cold to warm. In summer, the sea is colder than the land, so warm, moist sea air blows inland, bringing rain. In winter, the land is colder than the sea, so dry air blows from land to sea.

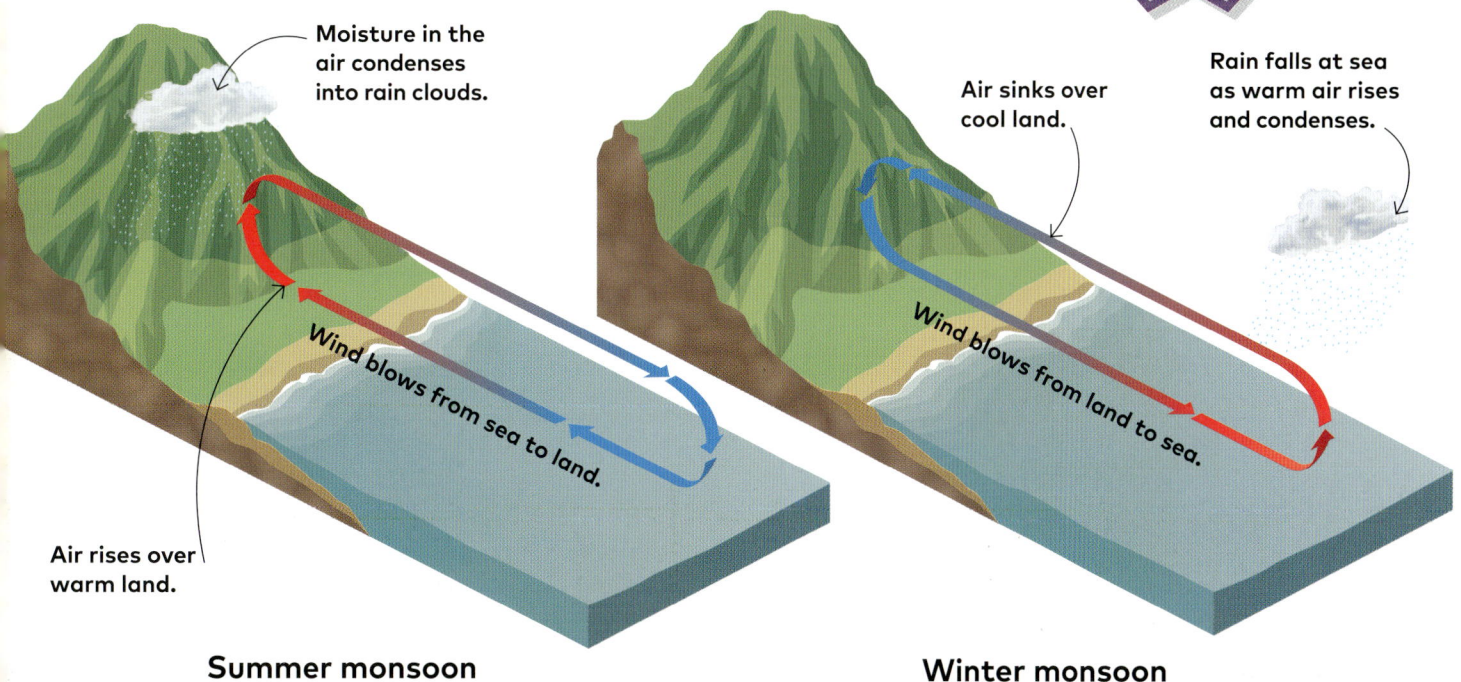

MAWSYNRAM, INDIA, IS THE WORLD'S WETTEST PLACE, WITH NEARLY 1,200 CM (470 IN) OF RAIN EACH YEAR.

Moisture in the air condenses into rain clouds.

Air rises over warm land.

Wind blows from sea to land.

Summer monsoon

Air sinks over cool land.

Rain falls at sea as warm air rises and condenses.

Wind blows from land to sea.

Winter monsoon

Does snow always melt?

Not everywhere. In most places that get winter snowfalls, all the snow eventually melts as the seasons change. But on high mountains and in polar regions, some places are always under snow. It never gets warm enough for all the snow to melt.

IF THE FULL ANTARCTIC ICE SHEET MELTED, GLOBAL SEA LEVELS WOULD RISE BY 60 M (200 FT).

What is a snowline?

Above a certain height, called the snowline, mountains are always covered by snow. The snowline is 5,000 m (16,400 ft) in the tropics and gets lower as you move nearer to the poles. At the poles, the snowline is at sea level.

Long-lasting snow

Most parts of Antarctica get very little snow, especially away from the coast. Any snow that does fall usually lasts a long time.

Reflecting sunlight

White snow reflects most of the sunlight that falls on it, instead of absorbing its energy. This stops the snow from melting and helps to keep the planet cool.

How do glaciers and ice sheets form?

Glaciers and ice sheets form where more snow falls each year than melts. Over thousands of years, layers of snow build up, squashing the snow beneath into thick ice. The Greenland and Antarctic ice sheets are about 3.2 km (2 miles) deep.

Greenland ice sheet

Forever freezing

Even in the warmest parts of Antarctica, the temperature rarely gets much above freezing in summer. In the coldest parts, winter temperatures are lower than -60°C (-76°F).

QUICK QUIZ

1. What are the winter temperatures in the coldest parts of Antarctica?
   a) Less than -60°C (-76°F)
   b) Less than -50°C (-58°F)
   c) Less than -30°C (-22°F)

2. How do glaciers and ice sheets form?
   a) From fresh snow that freezes immediately to thick ice.
   b) From the slow build-up of layers of snow.
   c) From ice on the surface of the ocean.

See pages 132–133 for the answers

Why is it cold in the mountains?

High mountaintops tend to be bone-shiveringly chilly because the Sun heats the atmosphere of the Earth, and the air expands and cools the higher up you go. For every 1,000 m (3,300 ft) you climb, the air cools by about 6°C (43°F), and sometimes as much as 10°C (50°F).

Cloudy summit

Tall mountains often have their peaks in clouds, which form when moist air travels over the top. Lower peaks may be misty or foggy.

Blustery heights

As well as the temperature falling as you get higher up a mountain, the wind tends to blow more strongly, too.

Why is one mountainside drier?

Moist air pushed up over a mountain cools and forms clouds, which drop rain (or snow) on the wind-facing side. There is little precipitation on the sheltered side, which is drier and is said to have a "rain shadow".

Rain shadow

Dry air flows down sheltered slope

Clouds form and rain falls

Moist air flows up wind-facing slope

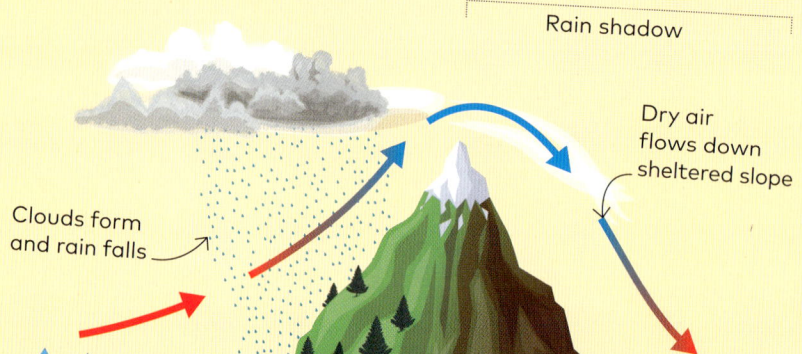

Snowy peaks

The summits of some mountains never get warm enough for the snow to melt, so they are always capped with snow.

TRUE OR FALSE?

1. More rain falls on windy mountain slopes than sheltered ones.

2. Mountaintops are warmer than the surrounding lowlands.

3. Wind chill makes it feel warmer than the actual air temperature.

See pages 132–133 for the answers

What is wind chill?

Wind chill is the way the wind takes heat away from your body as it blows over you. It can make the weather feel much colder than the actual air temperature.

Polar easterlies

Winds around Earth's polar regions are called polar easterlies. These dry, often bitterly cold winds begin blowing away from the poles, but then swerve from east to west.

Trade winds

Tropical trade winds blow towards the Equator, from east to west. Sailing ships once relied on these steady winds to blow them quickly over the ocean as they carried traded goods.

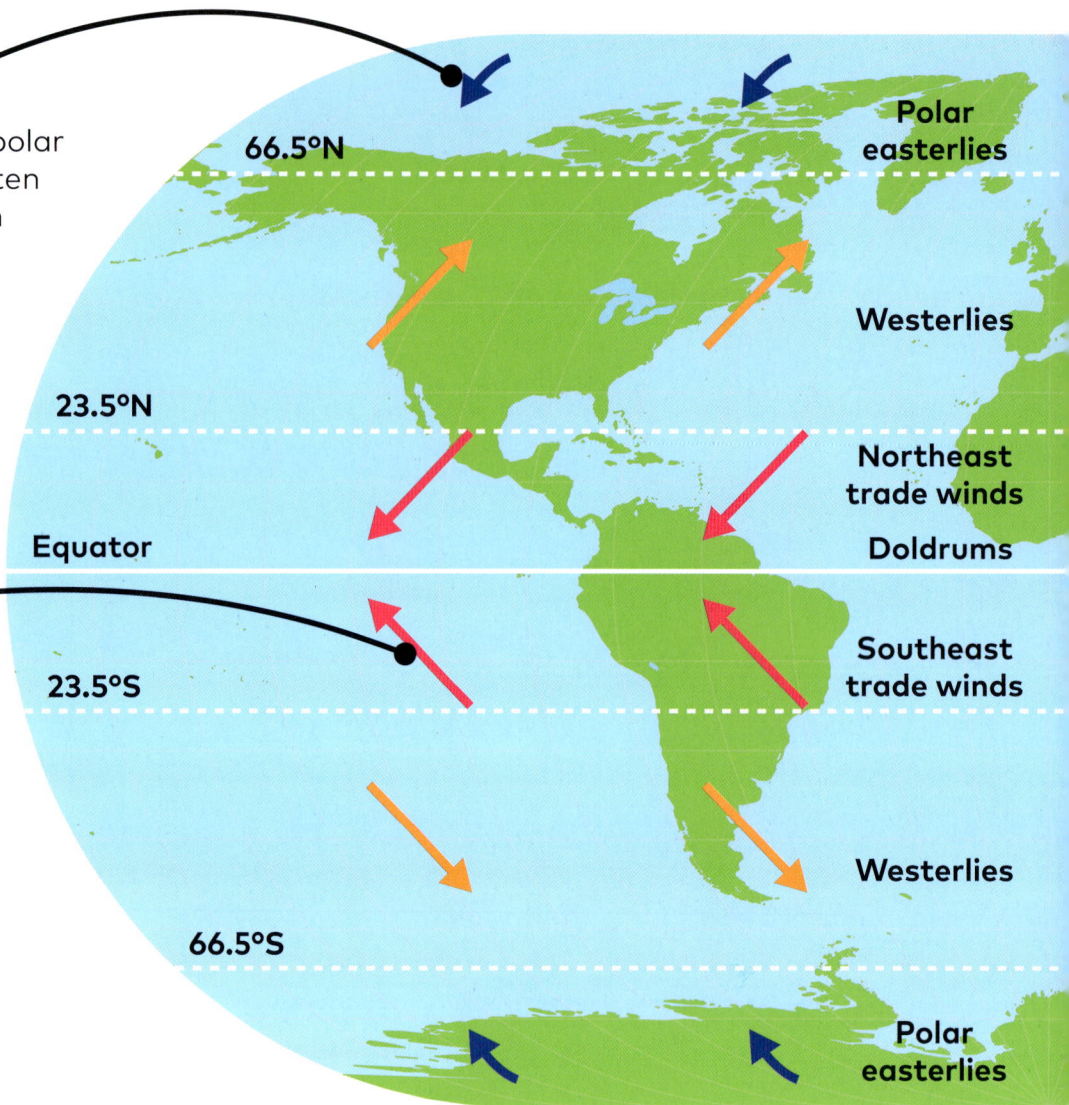

66.5°N

23.5°N

Equator

23.5°S

66.5°S

Polar easterlies

Westerlies

Northeast trade winds

Doldrums

Southeast trade winds

Westerlies

Polar easterlies

Are winds different across the world?

Yes, they can be. The major winds that blow most of the time, called prevailing winds, blow from different directions in different parts of the world. Local winds can whip up at different times, but prevailing winds blow consistently over huge swathes of our planet's surface.

PLANES CUT FLIGHT TIMES BY RIDING FAST, HIGH-LEVEL WINDS CALLED **JET STREAMS.**

QUICK QUIZ

1. Which way do trade winds blow?
   a) East to west
   b) North to south
   c) West to east

2. Which of these is not a prevailing wind?
   a) Polar easterly
   b) Sea breeze
   c) Westerly

See pages 132–133 for the answers

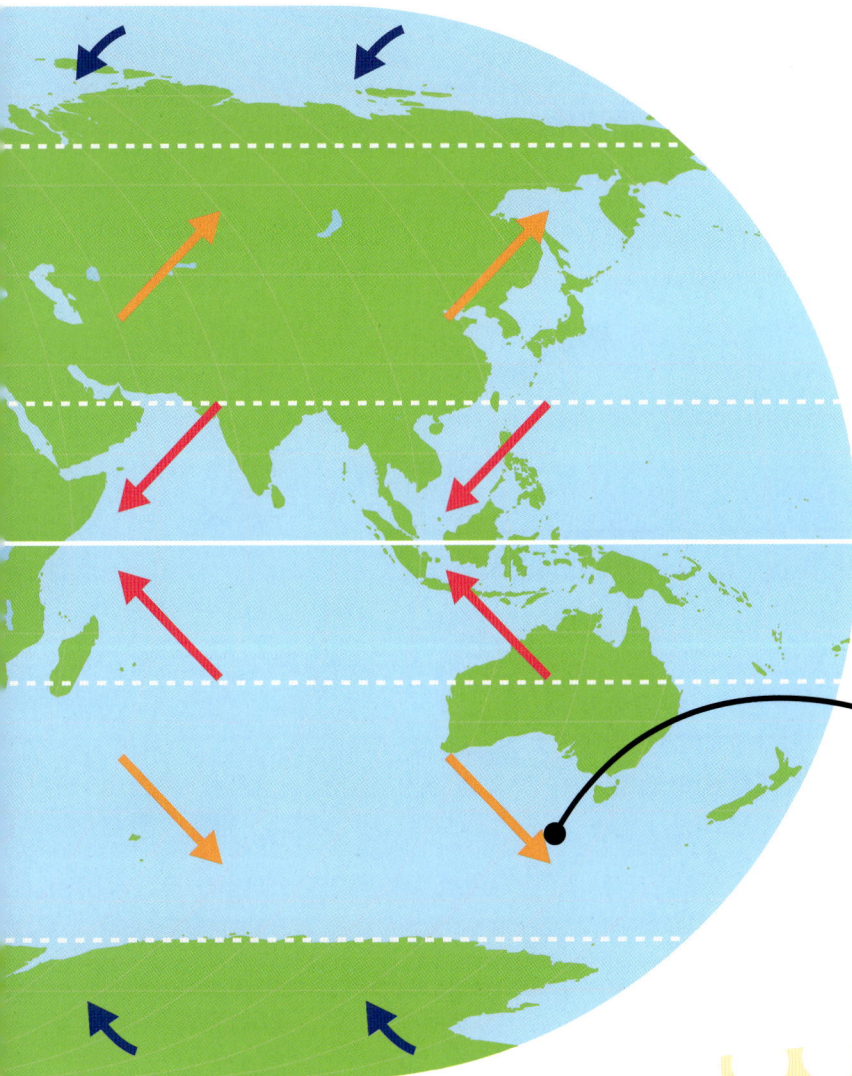

Westerlies

In the mild-weather regions between the poles and the tropics, there are warm, moist, westerly winds. These winds blow away from the Equator, then veer west to east.

What are the doldrums?

The doldrums is a low-pressure area at the Equator, between the trade winds, where there is little surface wind. The Sun's intense heat makes air flow straight upwards rather than horizontally. Sailors used to dread getting stranded in the doldrums. With no wind to stir their sails, they could be stuck for weeks!

How does Earth's spin affect winds?

Earth's rotation stops prevailing winds from blowing in straight lines. Instead, their paths are bent by Earth's spin. In the northern hemisphere, winds are flung to the right, and in the southern hemisphere, to the left.

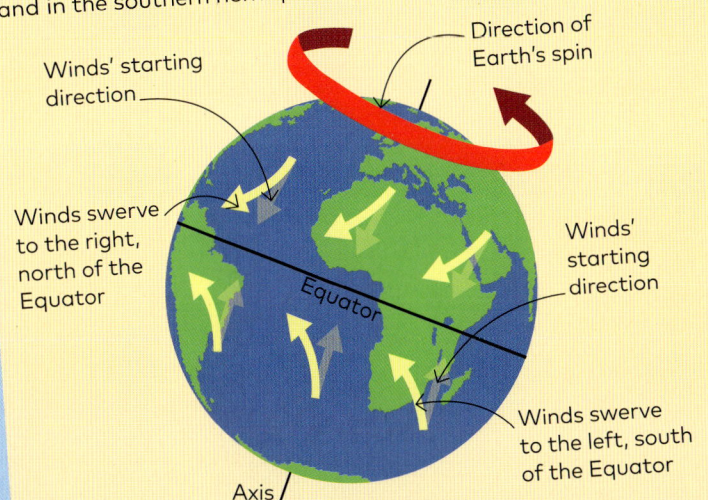

Direction of Earth's spin

Winds' starting direction

Winds swerve to the right, north of the Equator

Equator

Winds' starting direction

Winds swerve to the left, south of the Equator

Axis

Extreme weather

The weather brings the rain and sunshine that sustain life on Earth, but it can be highly destructive, too. Swirling tornadoes, battering hurricanes, unstoppable floods, and raging wildfires cause devastation, while billowing dust storms and blizzards bring daily life to a standstill.

What are tropical cyclones?

Tropical cyclones are Earth's most violent weather events. They brew up over warm seas either side of the Equator and often travel vast distances over the ocean. If they hit land, they can cause chaos and devastation. Depending on where they happen, they can be called cyclones, hurricanes, or typhoons.

THE LARGEST TROPICAL CYCLONE, TYPHOON TIP, WAS 2,200 KM (1,380 MILES) WIDE – NEARLY HALF AS WIDE AS THE USA!

What damage can a tropical cyclone do?

Powerful winds can blow down buildings, and lashing rain may cause flooding. A storm surge of up to 10 m (30 ft) and huge waves whipped up by the wind can race far inland, sweeping away everything in their path.

Monster storm
Tropical cyclones are typically about 480 km (300 miles) wide, but they can vary greatly in size. The cloudless eye is usually 30–60 km (20–40 miles) across.

How do tropical cyclones move?

Tropical cyclones move from east to west on the trade winds, then veer north above the Equator and south below it. When they move over cooler land or sea, cyclones lose energy, weaken, and eventually die away.

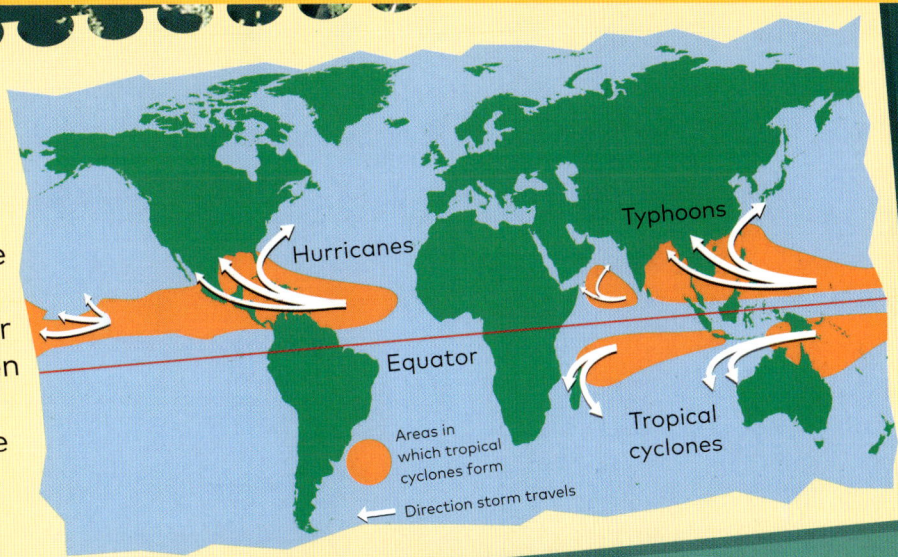

Hurricanes

Typhoons

Equator

Areas in which tropical cyclones form

Tropical cyclones

← Direction storm travels

Swirling clouds

Earth's rotation causes the whole storm to spin. Tropical cyclones swirl anticlockwise in the northern hemisphere and clockwise in the southern hemisphere.

Calm eye

The calm centre is called the eye. The strongest winds blow around the eye, at up to 400 kph (250 mph). Low pressure in the eye causes the sea level below to rise, creating a "storm surge".

Flat top

A thick, circular roof of cirrus cloud forms the top of the storm. Beneath it are rings of huge thunderstorms that can shed a month's worth of rain in a few hours.

TRUE OR FALSE?

1. The fastest winds in a tropical cyclone are at the storm's outer edge.

2. Trade winds move tropical cyclones east to west over the ocean.

3. All tropical cyclones are called hurricanes.

See pages 132–133 for the answers

How do tropical cyclones form?

A tropical cyclone may form if the ocean's surface reaches 27°C (80°F). The high temperature causes vast amounts of water to evaporate from the sea, forming rings of storm clouds around an area of low pressure. The storm system grows in size and power until it becomes a swirling tropical cyclone.

MOST TROPICAL CYCLONES DIE OUT AFTER ABOUT A WEEK, BUT **CYCLONE FREDDY LASTED OVER 5 WEEKS** IN 2023!

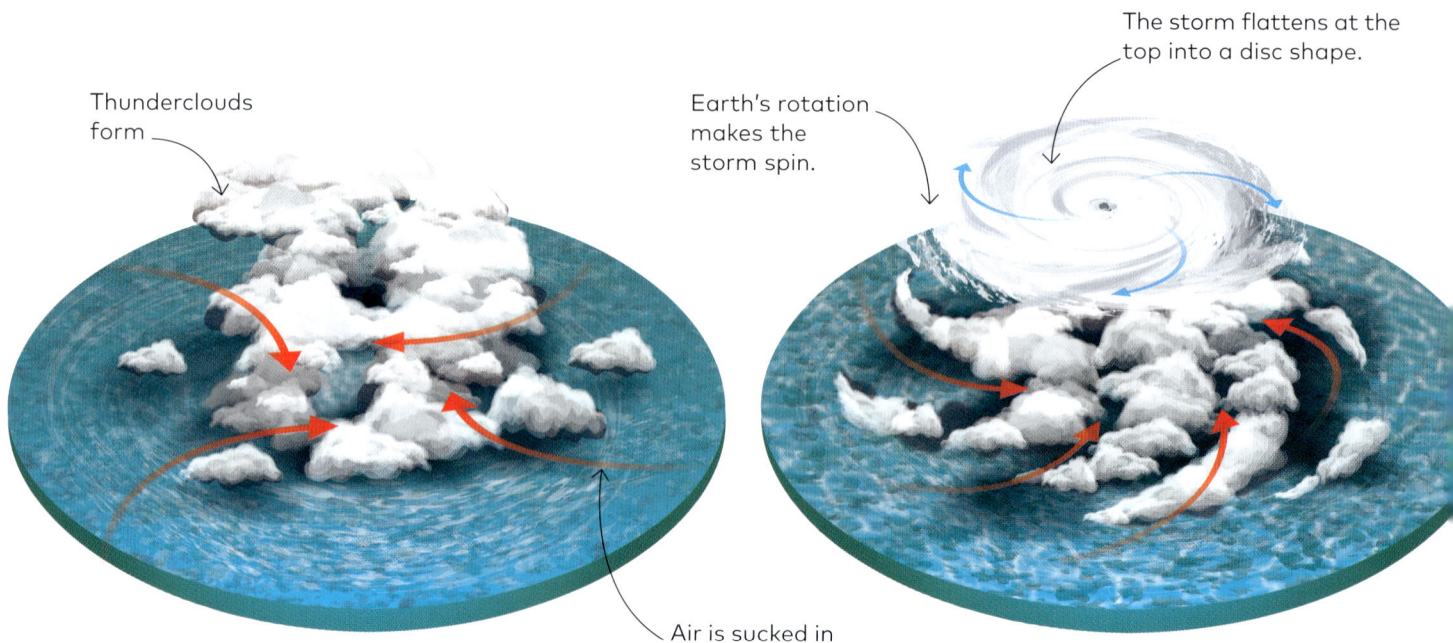

Thunderclouds form

Air is sucked in

1. Thunderclouds

Warm, moist air above the ocean spirals upwards. It cools and condenses to form thick bands of billowing cumulonimbus thunderclouds. Winds blow as air is sucked in by low pressure.

The storm flattens at the top into a disc shape.

Earth's rotation makes the storm spin.

2. Tropical storm

The winds get stronger as more cool air rushes in to replace the rising warm, moist air. The storm system begins to spin. When the winds blow faster than 63 kph (39 mph) it is known as a tropical storm.

What are hurricane hunters?

Hurricane hunters are planes packed with scientific instruments that fly right into the heart of tropical cyclones to take measurements. The information they collect helps forecasters predict the path of the storm, so people can be evacuated to safety before it strikes.

See pages 132–133 for the answers

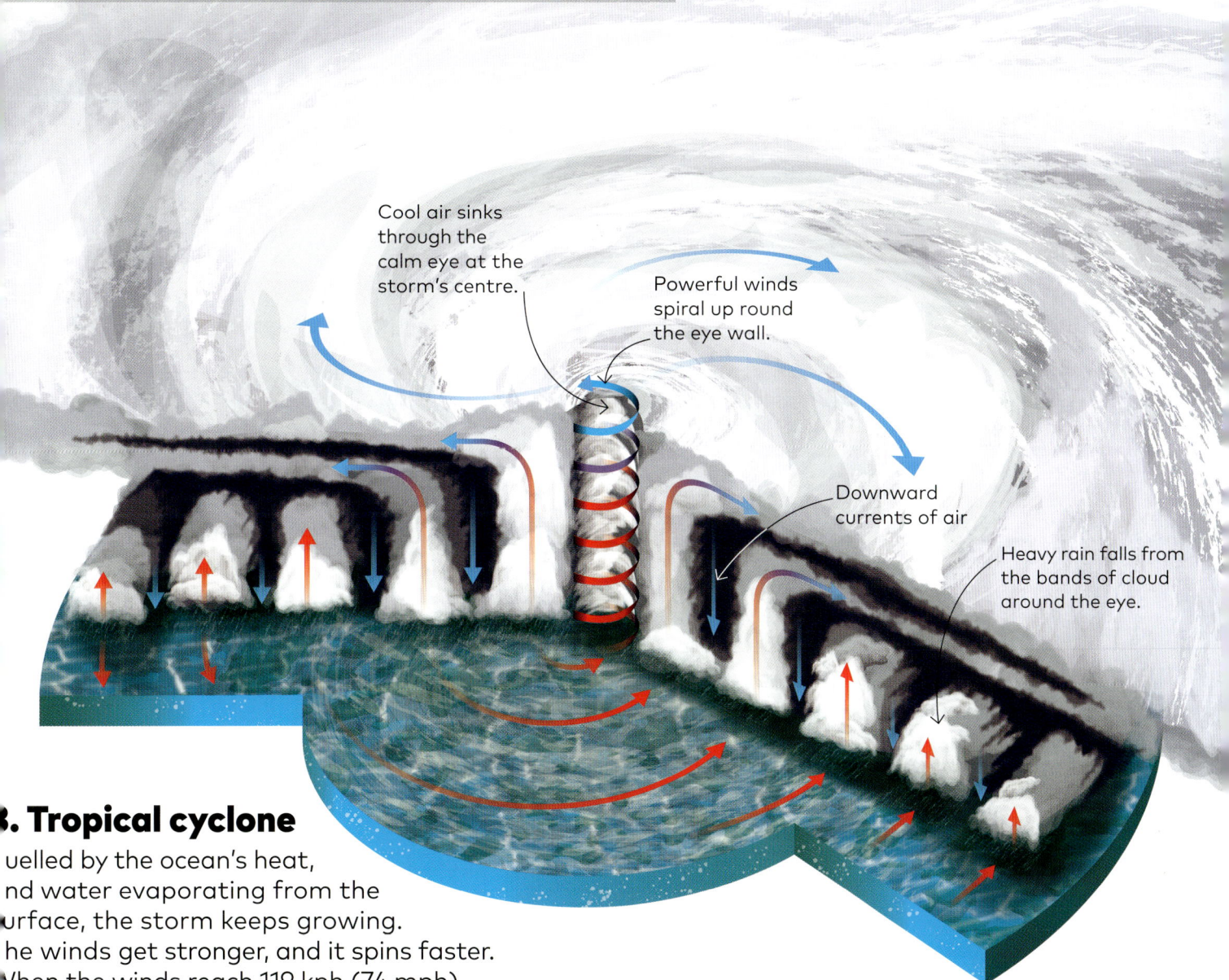

QUICK QUIZ

1. Which is the most powerful: tropical storm, tropical cyclone, or thunderstorm?

2. How warm does the ocean have to be for tropical cyclones to form?

Cool air sinks through the calm eye at the storm's centre.

Powerful winds spiral up round the eye wall.

Downward currents of air

Heavy rain falls from the bands of cloud around the eye.

3. Tropical cyclone

uelled by the ocean's heat, nd water evaporating from the urface, the storm keeps growing. he winds get stronger, and it spins faster. When the winds reach 119 kph (74 mph), is called a tropical cyclone.

What are tornadoes?

Tornadoes, sometimes called twisters, are ferocious columns of spinning air. They can produce the fastest winds on Earth, roaring up to 480 kph (300 mph). Tornadoes are born from thunderstorms. When they touch the ground, they cause devastation. Tornadoes are most common in the USA, but they happen elsewhere, too.

THE **MOST TORNADOES** RECORDED IN 24 HOURS IS **175**, ON 27–28 APRIL 2011 IN THE USA.

Are there different tornado types?

Rope tornado
Long, thin rope tornadoes snake between the sky and ground. Many tornadoes begin like this, then grow into bigger twisters. Others end as rope tornadoes as they dwindle away.

Waterspout
When a tornado develops over the sea, it is called a waterspout. It is mostly cloud, mist, and sea spray. Waterspouts are usually gentler and longer-lasting than land tornadoes.

Twin tornadoes
Sometimes two or more funnels may drop down from the same "parent" supercell cloud. If one is weaker than the other, it usually rotates around its sister tornado.

Touchdown

Where it touches the ground, the tornado sucks up dust and debris. As it moves over the landscape, it can pick up heavy objects, too, from cows to cars, and drop them far away. The dust, and condensing water vapour from the air, give the tornado its dark colour.

Monster cloud

The most destructive tornadoes come from vast thunderclouds called supercells. These clouds cause severe weather, including strong winds, hail, and heavy rain. Only some supercells make tornadoes.

Spinning column

Low pressure at the centre of the whirling funnel of air pulls in the surrounding air. The air is drawn upwards and starts spinning at tremendous speed.

PICTURE QUIZ

What do you think this type of monster cloud is called?
a) A tornado
b) A supercell
c) Tropical cyclone

See pages 132–133 for the answers

Why do supercells form tornadoes?

Inside a supercell cloud is a spinning updraught of warm, moist air. When cold downdraughts push the spinning air down, a "funnel" pokes out from the base of the cloud. If the funnel touches the ground, it becomes a tornado.

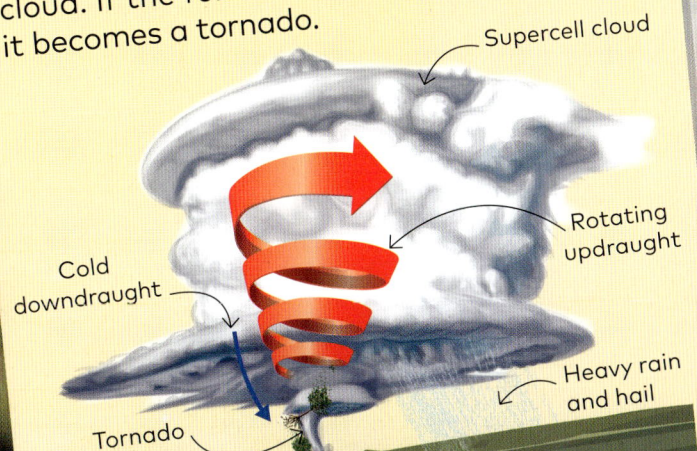

Supercell cloud

Rotating updraught

Cold downdraught

Heavy rain and hail

Tornado

What causes a heatwave?

Heatwaves are periods of unusually hot weather. They sometimes cause record-breaking temperatures. Heatwaves occur in summer when temperatures are the highest. They begin when high pressure builds up in the atmosphere, trapping warm air near the ground. A "heat dome" may form, making the heatwave much worse.

Heat dome

A high-pressure area forms as air sinks through the atmosphere. The sinking air acts like the lid of a cooking pot, trapping warm air beneath it as a heat dome.

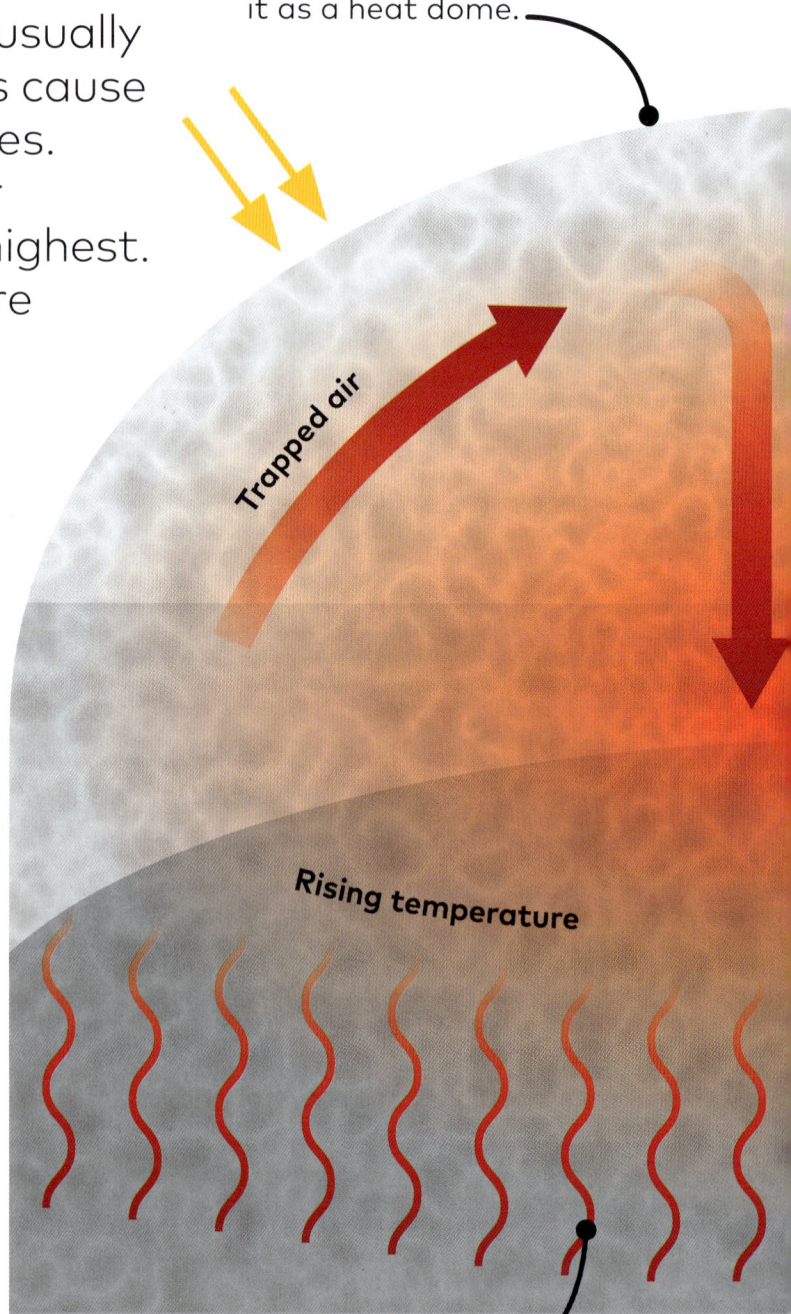

Trapped air

Rising temperature

What is the hottest temperature recorded on Earth?

The hottest-known temperature is 56.7°C (134°F), measured on 10 July 1913 in the desert of Death Valley, USA. Global warming might see this record smashed in the future. Heatwaves are becoming more common, lasting longer, and getting hotter.

Getting hotter

As the warm air sinks, it gets more compressed (squashed) and hotter. Temperatures at ground level keep rising. People, animals, and plants suffer in the heat.

High pressure

Trapped air

Calm, dry, summer weather warms the air. The warm air would normally rise and cool, but as it tries to rise, the high pressure above pushes it back down towards the ground.

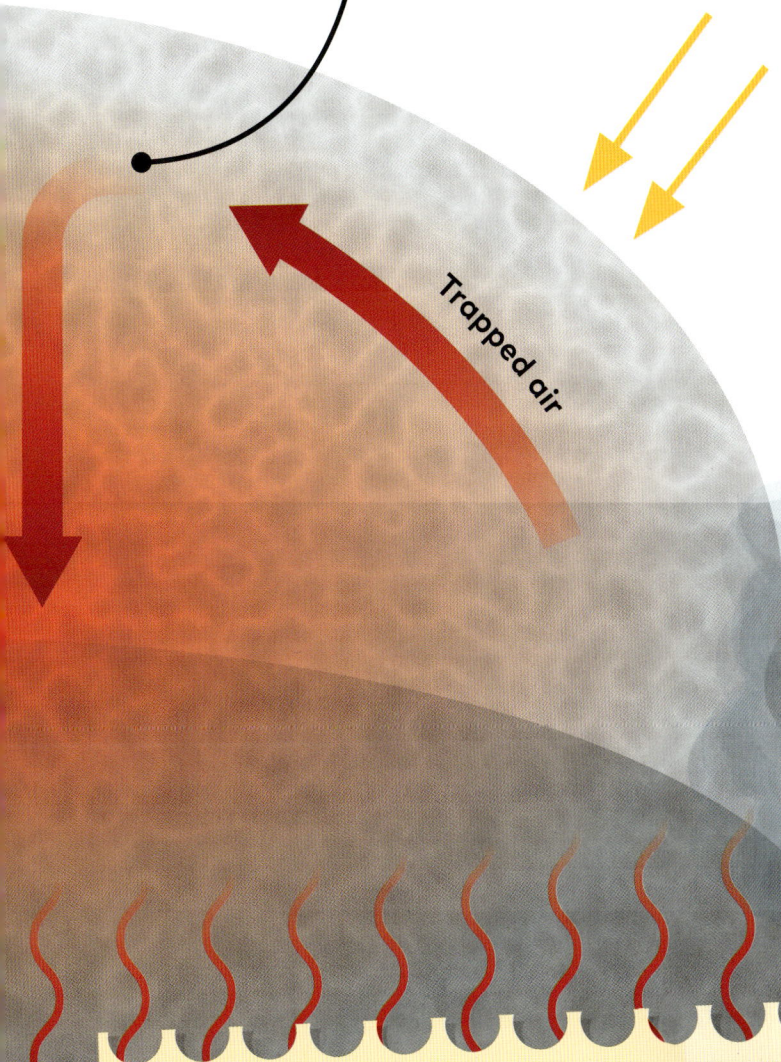

Trapped air

QUICK QUIZ

1. What causes a heatwave – high pressure or low pressure?

2. Are heatwaves becoming more common or less common?

3. What do heatwaves often end with?

See pages 132–133 for the answers

Cloud block

The heat dome prevents clouds from forming. Since the warm air inside the dome can't rise enough to cool, no clouds form within the dome, and there is little wind.

How does a heatwave end?

When high pressure parks itself over an area, it may stay for days or weeks. But the high pressure eventually weakens, heat begins to escape, and the air gradually cools. A heatwave often ends with thunderstorms as it is replaced by a cold front.

Can rain put out a wildfire?

Rain can help extinguish flames. It can also dampen the fire's fuel, making it harder for the fire to spread. It is unlikely to put out big, intense blazes on its own – that's where specially trained firefighters come in.

Wind assistance

Strong winds spread wildfires quickly and blow embers up to 40 km (25 miles) away, where they may start new fires. If the wind changes direction, so does the fire.

Forest furnace

In forest fires, the temperature on the forest floor can reach up to 1,200°C (2,190°F). The air heats up so much that things ahead of the fire may also burst into flames.

What causes a wildfire?

Wildfires can race over the countryside, consuming everything in their path. They happen when there is a long period of hot, dry weather and plenty of parched vegetation. Some fizzle out, but others – fanned by the wind – become raging infernos. These fires are difficult to put out once ablaze.

Fuelling the flames

A wildfire needs dry plant matter, such as leaves, grass, and branches, to fuel its burning. The more of this material there is, the longer the fire burns.

1. What does a wildfire need to keep burning?

2. When do wildfires usually happen?

3. Why will global warming make wildfires more likely?

See pages 132–133 for the answers

WILDFIRES TRAVEL FAST – AT ABOUT 20 KPH (12 MPH) OVER GRASSLAND AND OVER 10 KPH (6 MPH) THROUGH FORESTS.

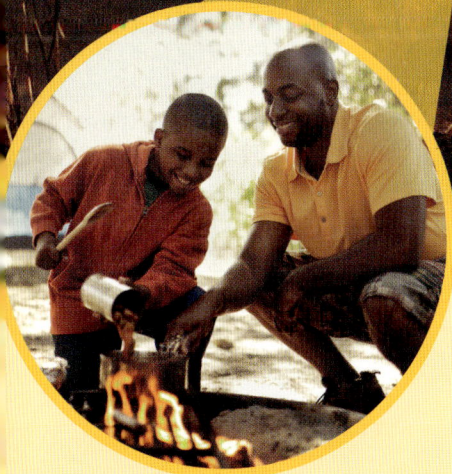

Is global warming causing more wildfires?

Yes, heatwaves are happening more often, making wildfires more common. In Australia, catastrophic fires raged in January 2020, after the country's hottest and driest year ever. Lightning is likely to increase with global warming, causing further fires.

Fire starters

Sometimes a lightning strike or the Sun's heat ignites the landscape. But many wildfires are started by people – a spark from a campfire may be all it takes.

Lumpy clump

Some of the biggest hailstones form when smaller hailstones collide and freeze together into one giant clump. The lumpy surface may even have horns or spikes of ice.

Giant hailstones

Big hailstones form when there are strong air currents inside storm clouds. Stronger air currents keep the hailstones aloft for longer, giving them time to form more layers.

How big are hailstones?

Hailstones can be as small as peas or as large as golf balls or even grapefruits! These hard ice pellets form in cumulonimbus storm clouds. Layers of ice then build up around the hailstones as they swirl around inside the cloud, until they get too heavy and fall to the ground.

Ice onion

This slice through a large hailstone shows how it is made up of layers, like an onion. The layers of ice may be clear, cloudy, or white if they contain trapped air bubbles.

What damage can hailstones cause?

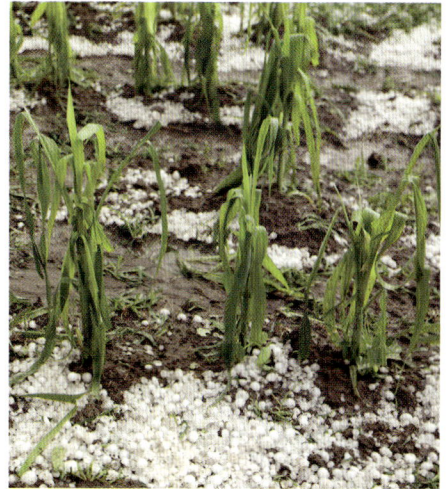

Large, heavy hailstones can destroy crops, dent cars, smash windows, and injure animals and people. It's not only their size that makes them destructive, but their speed, too – the biggest hailstones can fall faster than 160 kph (100 mph)!

How do hailstones form?

currents carry raindrops upwards inside storm clouds, here freezing temperatures turn the drops to ice. The young hailstones grow, layer by layer, as they collide with quid water drops that freeze onto their surface.

Hail grows as it swirls through the cloud.

Hail falls if it becomes too heavy, or the upward air currents weaken.

Cumulonimbus cloud

Raindrops freeze as they are carried upwards.

Warm air current

Cold air current

TRUE OR FALSE?

1. Hail forms in cumulus clouds.

2. Hailstones are made up of layers of ice.

3. All hailstones are the size of peas.

4. Hailstones can damage cars, windows, and crops.

See pages 132–133 for the answers

How does a dust storm form?

Dust storms happen when strong wind blows over dry places with few plants and trees. The wind breaks up the dusty layer covering the soil and lifts huge amounts of dust into the air. A rolling cloud of dust sweeps over the land, travelling hundreds or even thousands of kilometres.

AT ANY ONE TIME, THERE IS ABOUT 20 BILLION KG (44 BILLION LB) OF DUST IN THE AIR.

Wall of dust
The billowing dust cloud engulfs everything in its path. Because it moves so fast, a dust storm can arrive suddenly, and take people by surprise.

What are dust devils?
Dust devils are spinning columns of dust-filled air. These whirlwinds form in dry places. When the Sun heats the ground, the warmed air above it starts to spiral upwards, carrying dust with it. Dust devils usually only last a few minutes.

Dust danger

he dust blocks out the Sun,
making it difficult to see. It can
ause breathing problems, destroy
rops, ruin machinery, ground
ircraft, and even spread disease.

Huge cloud

A dust storm can be several
ilometres long and thousands
f metres high. After it is over,
ust particles may still hang in
he air for days or even weeks.

TRUE OR FALSE?

1. Sand grains are lighter than dust particles.

2. Dust storms form in wet regions with plenty of plant cover.

3. Dust devils start on the ground, not in the sky like tornadoes.

See pages 132–133 for the answers

Are sandstorms and dust storms the same?

Sandstorms are formed from wind-blown sand
grains, not dust particles. Unlike dust storms, they
only happen in desert regions. They also tend to happen
on a smaller scale – sand is heavier than dust, so it is
harder for the wind to keep the sand airborne.

Can volcanoes change the weather?

Volcanic eruptions can alter the weather both locally and globally. Near an eruption there is often rain, thunder, and lightning. Material hurled into the atmosphere and spread by winds can travel right around the world, and even cool the planet.

Can volcanoes affect the climate?

Along with dust, volcanoes belch sulphur dioxide gas into the stratosphere. The gas combines with water to form a haze of sulphuric acid droplets. The dust and haze block out sunlight, cooling the climate until they fall to Earth.

Lightning bolts

Particles of water vapour, ash, and dust in the volcanic plume collide and become charged with electricity. The build-up of charge is released as lightning.

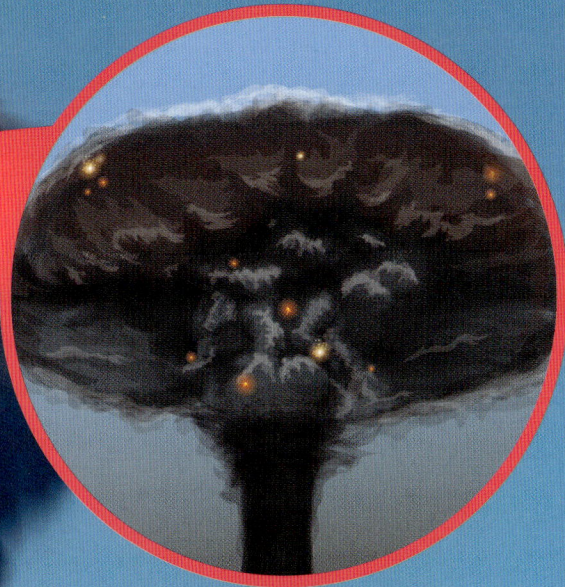

Volcanic clouds

Water vapour condenses around airborne ash, creating vast, billowing clouds that may bring rain and thunderstorms.

Rising plume

The smoky column that rises from an erupting volcano is called a volcanic plume. It is a mixture of hot ash, dust, water vapour, and other gases and air.

Eyjafjallajökull, Iceland, April 2010

Do eruptions change the colour of the sky?

Volcanic ash, dust, and haze scatter the colours in sunlight more than usual. When the Sun is low in the sky, its light must pass through lots of these volcanic particles, giving us especially fiery sunsets and sunrises.

What causes floods?

Floods happen when bodies of water overflow and submerge land. Powerful flash floods happen very suddenly, when storms dump huge amounts of rain on a place in a very short time. Other, less-powerful floods happen when river levels build up more slowly as it rains heavily over many days.

JUST 60 CM (20 IN) OF FAST-MOVING WATER CAN FLOAT A CAR AWAY!

Much more rain falls than the ground can soak up.

Homes can be cut off by the water and people left stranded.

1. Downpour

A sudden, intense rainstorm sends water rushing down steep hill slopes into the river valley below. The river level quickly rises, and the river flows faster and more powerfully.

2. Flood

The river can't hold any more water and it bursts its banks, submerging fields, roads, and streets, and flooding buildings. A raging torrent carries away fallen trees, cars, and debris.

What are El Niño floods?

El Niño is a weather pattern. Every few years, Pacific Ocean trade winds weaken or change direction and blow east instead of west. This has a knock-on effect on weather around the world, especially in the tropics. Events such as floods and drought are more likely to happen in El Niño years.

QUICK QUIZ

1. What do we call a flood after a sudden burst of rain?
   a) A tidal wave
   b) A flash flood
   c) A storm surge

2. El Niño is caused by changes to which winds?
   a) Polar easterlies
   b) Atlantic Ocean trade winds
   c) Pacific Ocean trade winds

See pages 132–133 for the answers

There may be serious damage to buildings.

3. Aftermath
The rain stops, but debris piled up against bridges may block the river's flow and cause more water to spill from the river. The water level gradually falls, leaving widespread damage.

How can we protect against floods?
Raised banks, called levées, can allow rivers to carry more water, and dams can be built to control water flow. Flood barriers across river mouths stop ocean tides raising river levels and flooding inland areas. The Thames Barrier protects the city of London, UK, from floods.

What is a blizzard?

A heavy fall of snow blown by strong winds is called a snowstorm. This becomes a blizzard when the winds are blowing at speeds of at least 48 kph (30 mph). Snow can build up in deep mounds, called drifts.

Whiteout

When there's snow blowing in the sky and lying thick on the ground, you sometimes get a "whiteout". It becomes difficult to see things that are even half a metre away.

THE **GREAT BLIZZARD** OF 1888 IN THE USA LASTED FOR FOUR DAYS.

Snow squalls

When sudden, heavy snowfall comes with overpowering winds, it is called a snow squall. It's like a miniature blizzard! These storms are especially common around lakes in early winter months.

How do snowstorms affect daily life?

Snowstorms can cause havoc as roads, railway tracks, and airport runways get blocked with snow. Accidents become common around this time. Snowploughs are used to clear the roads, moving the snow into large piles. Drivers put tyre chains on cars for better grip, to avoid slipping.

QUICK QUIZ

1. Snow squalls are common in which months?

2. Which vehicles are used to push the snow off the roads?

See pages 132–133 for the answers

Ice storm

Winter storms often bring freezing rain that turns to ice as soon as it touches anything. These ice storms can be dangerous to aircraft and other vehicles. Roads become glazed with slippery ice, and trees and power lines often snap under the extra weight.

Living with weather

Plants and animals in the wild have evolved to live in every type of climate on Earth. Using our skills and knowledge, humans can do the same. We study the weather to learn how, when, and why it changes, so that we can be better prepared for whatever it brings.

Can people survive in any climate?

Humans can survive in most climates and places on Earth if they have the right clothing, shelter, and equipment, as well as reliable supplies of food and water. People have set up permanent villages, towns, and cities in some of the harshest conditions on the planet.

THE TEMPERATURE AT **OYMYAKON, RUSSIA**, ON 6 FEBRUARY 1933 WAS –67.7°C (–89.9°F)!

Bone dry

Arica, on the coast of Chile, gets almost no rain. The city gets its water from wells and the nearby Lluta River. Water is also taken from the sea, and the water's salt is removed to make it drinkable.

Aerial view of Arica

North America

Atlantic Ocean

Pacific Ocean

South America

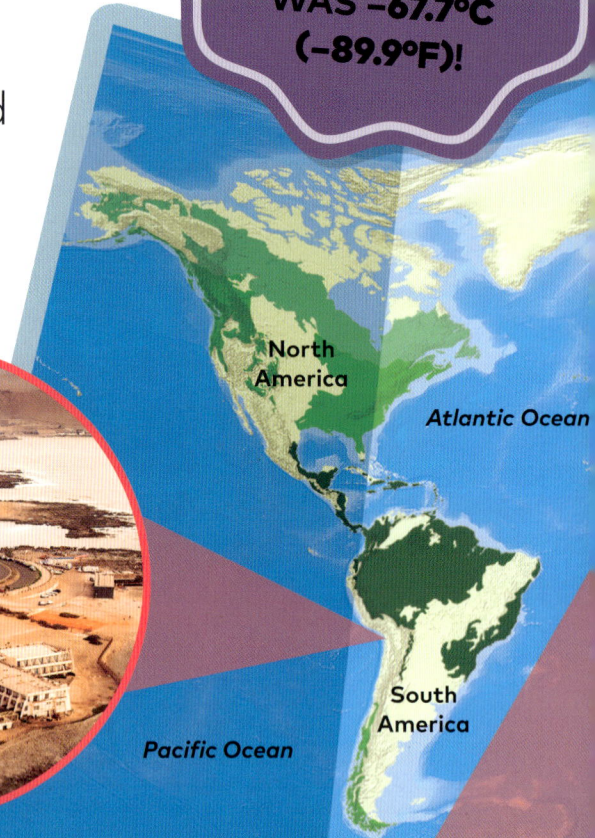

Does anyone live in Antarctica?

Yes, people do live there – but not permanently. There are research stations, such as Halley VI, where scientists go to study the continent. The scientists stay for part of the year, then others take their place. Antarctic buildings are highly insulated to keep out the cold.

Relentless heat

Houses in Araouane, Mali, at the Sahara Desert's edge, are made of dried mud. The mud walls keep homes from becoming too hot inside. Full-length clothes and headgear help to protect people from the Sun.

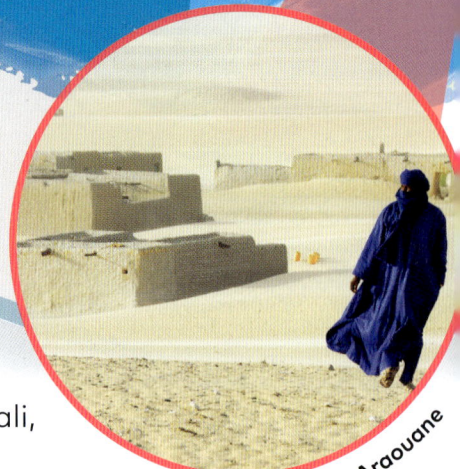

Desert village of Araouane

Below zero

In Oymyakon, Russia, it can be way below freezing for about six months of the year. Few crops grow because the ground freezes, so people in Oymyakon mostly eat fish and meat – sometimes frozen!

A man catching fish in Oymyakon

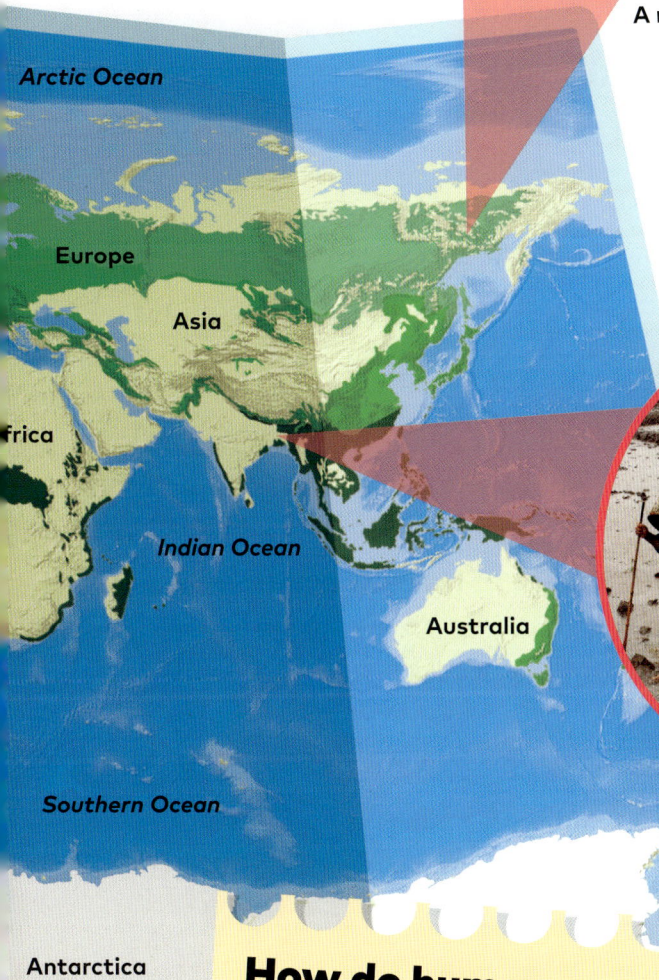

Arctic Ocean

Europe

Asia

Africa

Indian Ocean

Australia

Southern Ocean

Antarctica

Flooded fields in Mawsynram

Non-stop rain

Mawsynram, India, is the world's wettest place. People there keep the monsoon rain off with huge, hat-like rain shields called knups. Houses are soundproofed to quieten the noise of hammering rain.

How do human bodies cope with weather changes?

The temperature inside the human body must be 37°C (98.6°F) to keep the body healthy and working properly. When your body temperature changes, perhaps when you're out on a freezing morning or scorching afternoon, your brain acts to put things right.

2. Alert
Special cells in your skin sense the temperature change and alert your brain.

1. Trigger
The weather causes your body temperature to change.

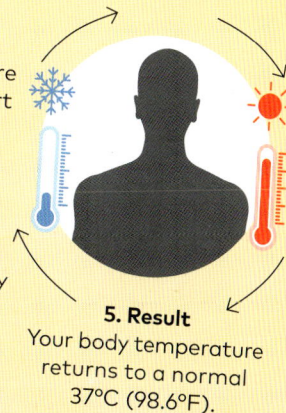

3. Control
Your brain asks: "Is my body temperature hotter or colder than 37°C (98.6°F)?"

4. Reaction
If you're too hot, your brain makes you sweat to cool down. If you're too cold, it makes you shiver to warm up.

5. Result
Your body temperature returns to a normal 37°C (98.6°F).

Extreme heat

A camel has woolly fur on its back to protect against the Sun and stop hot air reaching its skin. Heat-resistant pads on the camel's feet allow it to walk over the scorching desert sand.

Extreme cold

Like many polar animals, the polar bear has a fatty layer called blubber under its skin that traps body heat. Its fur lets the Sun's energy pass through to warm the bear's skin.

How do animals survive in extreme climates?

The answer is simple – they rely on their bodies. Animals living in harsh climates have evolved special features that help them cope with numbing cold, baking heat, shrivelling drought, and thin mountain air. When extreme weather comes, some animals remain active, but others hide away, resting until conditions improve.

Drought

In the dry season, the water-holding frog burrows into soil and sheds its skin, which hardens around it like a cocoon. The cocoon stops its body from drying out, keeping the frog alive until the rains return.

Thin air

The guanaco lives in high mountains. It has extra red blood cells to absorb as much oxygen as possible from the air it breathes. A super-sized heart pumps the oxygen-rich blood quickly around its body.

What is the toughest animal?

The tiny tardigrade is probably the toughest creature. Tardigrades are found in deserts, hot springs, glaciers, and on high mountaintops and volcanoes. They can survive freezing and boiling temperatures, crushing pressure, and even being zapped with ultraviolet radiation in airless space!

PICTURE QUIZ

What is the fatty layer under an emperor penguin's skin that keeps it warm on the Antarctic ice?

See pages 132–133 for the answers

How do plants grow in tough climates?

Plants have found many ways to overcome the challenges of living in tough climates. Cacti in hot, dry deserts store water to keep themselves alive between the rare desert downpours. And in very cold places, many plants hug the ground to avoid being dried out by fierce, icy winds.

Instead of leaves, the cactus has succulent stems with spines, which lose less moisture to the air.

Water tower

The pleated stem of the giant saguaro cactus can expand like a concertina to store water. Its tough, waxy skin stops the water from evaporating in the heat.

What is a desert bloom?

A desert bloom is when the desert becomes carpeted with flowers after a sudden rainstorm causes buried seeds to sprout. They grow quickly, flower, scatter their seeds, and die. The new seeds stay under the sand, waiting for rain.

Thirsty roots

The shallow, wide-spreading roots of a giant saguaro quickly soak up dew and any rain that falls. The cactus stores the water in its thick, fleshy stem.

Wind-cheater

Moss campion grows in low, rounded clumps in the Arctic and on high mountains. This shape keeps the plant out of the wind, while allowing it to get as much sunlight as possible.

Holding onto heat

Growing in low clumps helps moss campion to absorb heat released by the ground. This is important in winter, when the plant may be covered with snow.

PICTURE QUIZ

Is this a pile of rags in the desert or a plant?

See pages 132–133 for the answers

Does icy weather hurt plants?

Plants used to warmer weather can be damaged or even killed if ice forms in their cells during an unusually cold spell. But some plants, including the winter-flowering snowdrop, contain natural antifreeze chemicals that stop their cells icing up.

Temperature

A U-shaped thermometer records the highest and lowest temperatures each day. Liquid in the tube expands as the temperature rises, and shrinks as it falls, moving markers on the thermometer.

Air pressure

An aneroid barometer has a face like a clock. It contains a box that changes size when the air pressure rises or falls. As the box changes size, it moves a pointer around a dial.

Wind

A propeller anemometer produces an electric current as it spins in the wind. The current's strength shows the wind's speed, and the way the propeller points shows the wind's direction.

How do we measure the weather?

We can take measurements of the weather using a range of simple instruments. They help us record temperature, air pressure, wind speed and direction, humidity, and rainfall. Weather scientists also use electronic equipment to monitor the weather.

IN 1644, EVANGELISTA TORRICELLI MADE THE FIRST BAROMETER TO TAKE EARLY MEASUREMENTS OF AIR PRESSURE.

Humidity

A wet-and-dry hygrometer has two thermometers: one bulb is kept wet, the other dry. Evaporation from the wet bulb creates a difference in temperature between the two, which shows humidity.

Rainfall

A rain gauge has a funnel that channels rainwater into a collecting tube. The depth of water in the tube is recorded, then the tube is emptied and dried ready for the next day.

What were old weather instruments like?

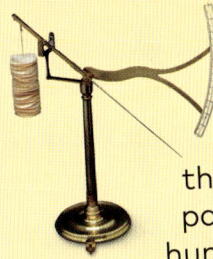

18th-century hygrometer
Unlike the modern hygrometer, this instrument used paper discs to measure humidity. In damp air, the discs absorbed water and weighed more, pulling the pointer up the scale.

17th-century thermometer
Looking like a candlestick, this 17th-century glass thermometer showed changes in temperature by the rise or fall of the coloured glass balls in the water-filled tubes.

What is a Stevenson screen?

To shade them from direct sunlight and keep out precipitation, weather instruments are often housed in white boxes called Stevenson screens. Slats in the sides allow air to flow freely around the instruments, while the white paint reflects the Sun's rays.

QUICK QUIZ

1. What did Evangelista Torricelli invent in 1644?
   a) Barometer
   b) Anemometer
   c) Thermometer

2. What do we call an instrument that measures humidity?
   a) Hydrometer
   b) Hydraulic meter
   c) Hygrometer

See pages 132–133 for the answers

Birds

When you see birds flying in a circle, a storm may be brewing. Birds detect changes in air pressure, so they fly round and round looking for a safe place to settle before bad weather arrives.

Seaweed

Kelp is sensitive to humidity. If you hang some kelp outside, and it dries and shrivels, the weather should be fine. When there's rain ahead, it absorbs moisture and feels damp.

Pine cones

When good weather is o the way, pine cones ope As humidity falls, the cone scales dry, stiffen and spread out. In wet weather, the scales swe and the cone shuts tight

Can nature predict the weather?

Before scientists found ways to measure and study conditions in the atmosphere, people used nature to forecast the weather. They scanned the sky and watched how animals and plants behaved for signs of change. Some of the signs hold true, but not all can be relied on.

What does a halo around the Moon mean?

A Moon halo can be a good predictor of rain, especially in summer. The halo appears when moonlight shines through high cirrostratus clouds, which often come before bad weather. The halo is caused by ice crystals in the clouds bending the light rays.

See pages 132–133 for the answers

QUICK QUIZ

1. Which of these is a sign of coming rain?
   a) A halo around the Moon
   b) Pine cones closing
   c) Both of the above

2. What does a red sky in the morning mean?
   a) Rain is likely
   b) Sunny weather is likely
   c) Nothing at all

Red sky

A red sky at sunset signals that high pressure is moving in, so the next day will be fine. A red sky at dawn means the fine weather has moved off and wet, low-pressure weather is coming.

Flowers

Insects that pollinate flowers tend not to fly if it's wet. Some flowers, like the scarlet pimpernel, shut when the sky darkens before rain. This keeps their pollen dry until the insects fly again.

Frogs

Expect rain when you see frogs out and about, and hear them croaking more than usual. Frogs need damp conditions, so they become more active when the humidity rises before rainfall.

Can science predict the weather?

Science can usually give us a very good idea of what the weather will be like. Data (information) is collected by orbiting satellites, high-flying aircraft and balloons, and radar and weather stations at Earth's surface. Computers use the data to work out what the weather is going to do.

Radar

A radar antenna sends out radio waves that reflect off clouds, rain, and snow. The "echoes" that bounce back allow weather scientists to track precipitation and storms.

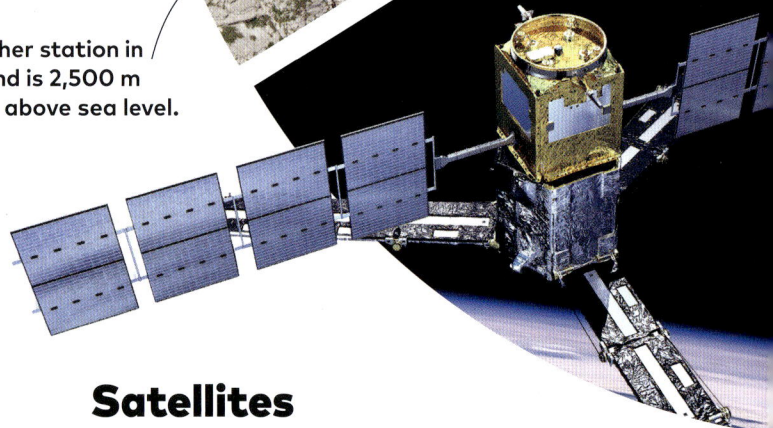

Weather stations

There are thousands of weather stations on land – from low plains to mountaintops – and thousands more on ships and buoys at sea. They monitor weather non-stop.

This weather station in Switzerland is 2,500 m (8,200 ft) above sea level.

WEATHER SUPERCOMPUTERS CAN DO TRILLIONS OF CALCULATIONS EVERY SECOND.

Satellites

Around 160 weather satellites take photos and scan Earth and its atmosphere with special sensors. These satellites make 80 million weather observations each day.

Computers

Powerful "supercomputers" use data about what the weather is like now to predict what the weather will probably do next. The results are sent to weather forecasters.

This supercomputer at the Japan Meteorological Agency helps to forecast heavy rains.

This balloon is being released by a scientist at a research station in Antarctica.

Balloons

Weather balloons carrying scientific instruments are sent high into the atmosphere. The instruments send measurements of the weather back to ground stations by radio.

Why do we need weather forecasts?

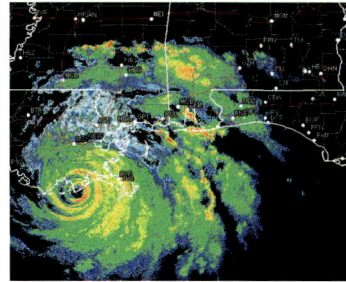

Bad weather is a hazard for transport. Weather forecasts help people know when it is safe to travel.

Farmers use weather forecasts to plan their crops and work out the best times to plant and harvest.

Advance warning of extreme weather events allows emergency services to be ready for action.

QUICK QUIZ

1. About how many weather satellites are circling Earth?

2. What does a radar antenna send out?

3. What happens to all the weather data that is collected?

See pages 132–133 for the answers

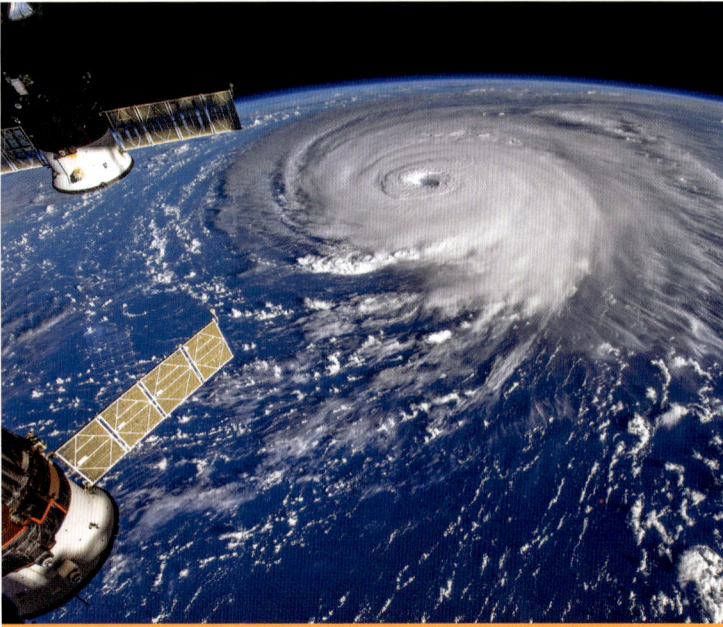

Studying weather

Meteorologists study how conditions in the atmosphere and at sea give different types of weather. The data (information) they collect allows them to forecast, or predict, what the weather will be like.

The climate of the polar regions is changing as the world warms up.

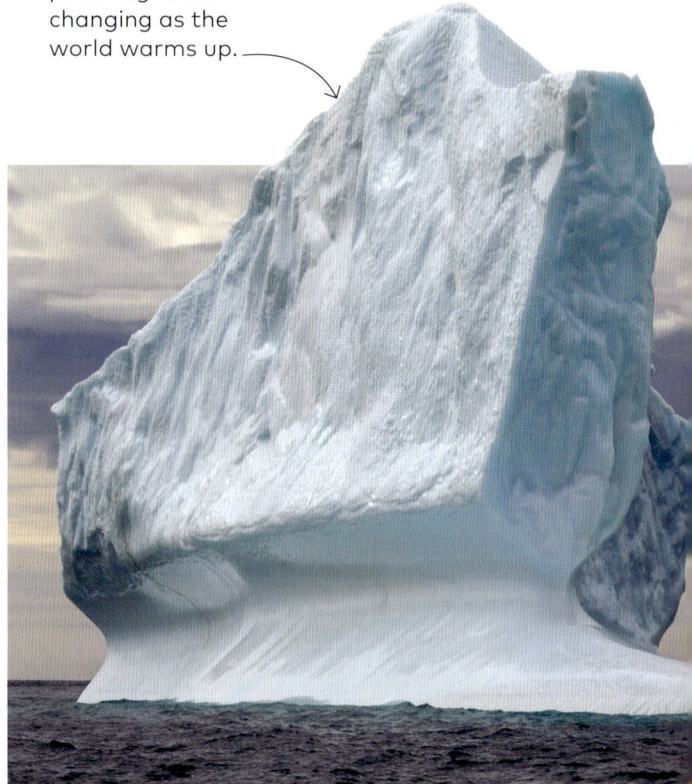

Investigating climate

Climatologists analyse weather patterns over months, years, or centuries. They try to understand what makes Earth's climate change, and how any changes might affect the planet in the future.

What do weather scientists do?

Over 2,000 years ago, Aristotle, a Greek philosopher, made the first study of the weather. Today, there are many types of weather scientist. They study the atmosphere – what it's like now and what it used to be like – and even investigate weather conditions in space.

POLLEN GRAINS FOUND IN **ANCIENT SEDIMENTS** TELL SCIENTISTS WHAT PLANTS LIVED IN PAST CLIMATES.

What are ice cores?

Ice cores are cylinder-shaped samples of ice that scientists drill out from immensely thick sheets of ancient ice. Bubbles of atmospheric gases trapped in the ice cores tell scientists what the climate was like long ago, when the ice formed.

The climate at the time of the dinosaurs was warmer and more humid.

QUICK QUIZ

1. What is another name for charged particles released by the Sun?
   a) Solar fire
   b) Solar wind
   c) Solar power

2. Why do scientists extract ice cores?
   a) To understand past climates
   b) To find fossils
   c) To cool down

See pages 132–133 for the answers

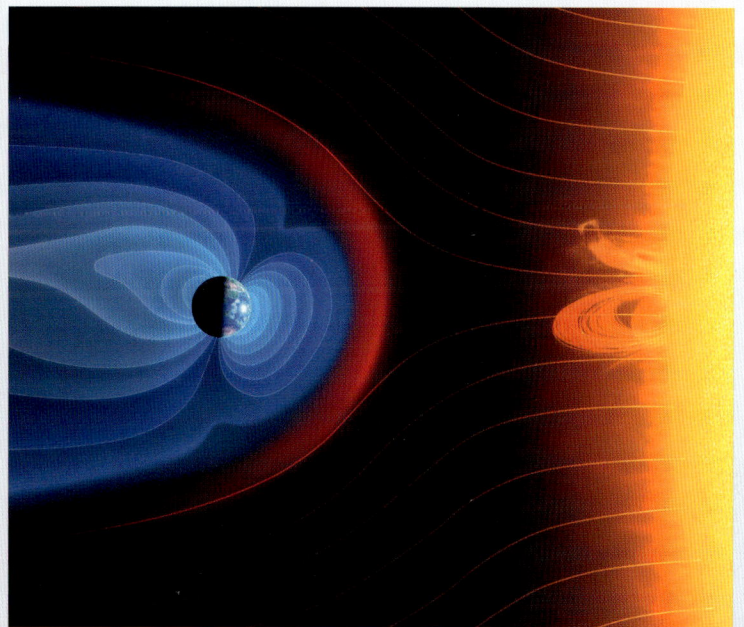

Uncovering the past

Paleoclimatologists investigate past climates ("Paleo" means ancient or old). They study rocks, ice sheets, and sediments, looking for clues that reveal the atmospheric conditions long ago.

Looking into space

Space weather scientists examine changes in the Sun to see how they might affect Earth. They are especially interested in the solar wind, a stream of charged particles that flows out from the Sun into space.

Climate change

Earth's climate never stays the same. It changes slowly over long periods of time, sometimes getting hotter, sometimes colder. Today, the climate is changing faster than usual because of human activities. Earth is getting warmer and extreme weather is happening more often. But there's plenty we can do to help the planet.

Does climate change happen naturally?

Yes, Earth's climate naturally see-saws between hot and cold periods. There are many causes of natural climate change. They include the shape of Earth's orbit, the angle of Earth's tilt, the movement of the plates of Earth's crust (outer shell), and changes in gases in the atmosphere.

Moving plates

As the plates of Earth's crust move over the planet, their motion affects the climate. It makes oceans grow and shrink, changing sea levels and currents.

The slabs of Earth's crust are called tectonic plates.

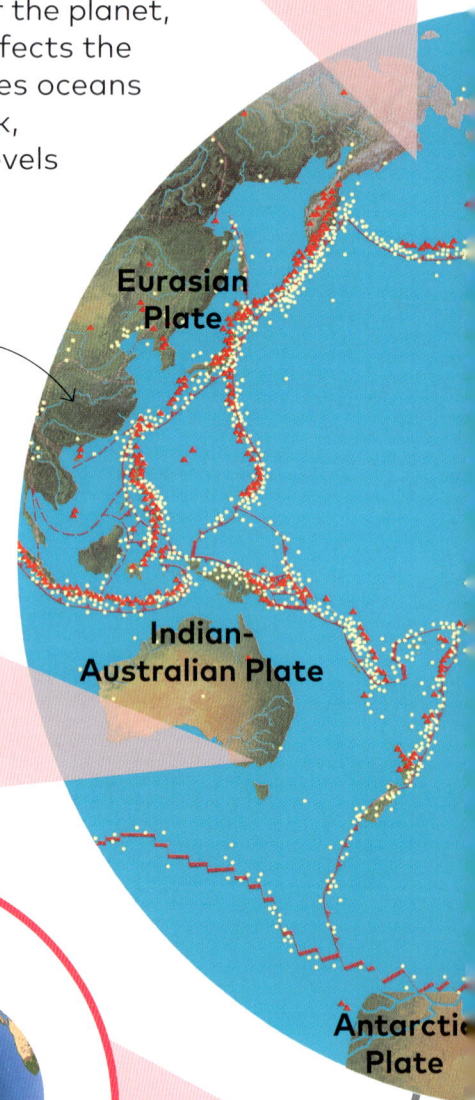

Eurasian Plate

Indian-Australian Plate

Antarctic Plate

Making mountains

Where Earth's crustal plates crunch together, high mountains are thrown up. New mountain ranges can affect the way the air circulates in the atmosphere.

Wobbling planet

Earth wobbles slightly on its axis as it spins, in a cycle that takes about 25,000 years. It makes seasons more extreme in one hemisphere and milder in the other.

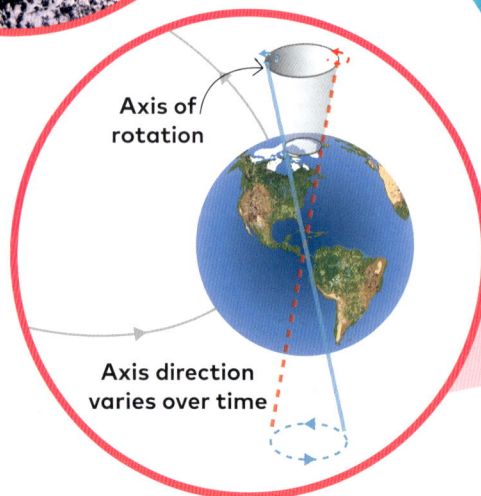

Axis of rotation

Axis direction varies over time

Changing tilt

Over about 40,000 years, Earth's tilt can vary between 22.1° and 24.5°. When the angle increases, summers get hotter and winters cooler. When it decreases, seasons are milder.

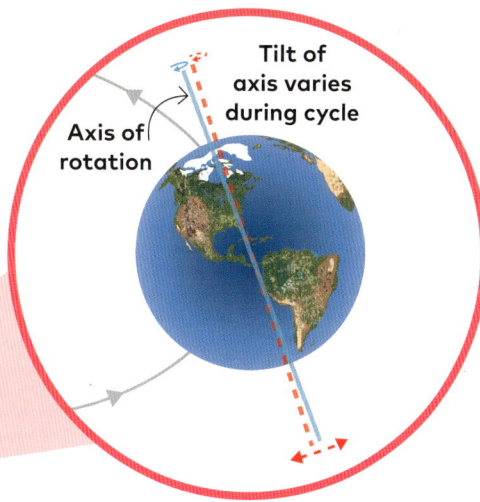

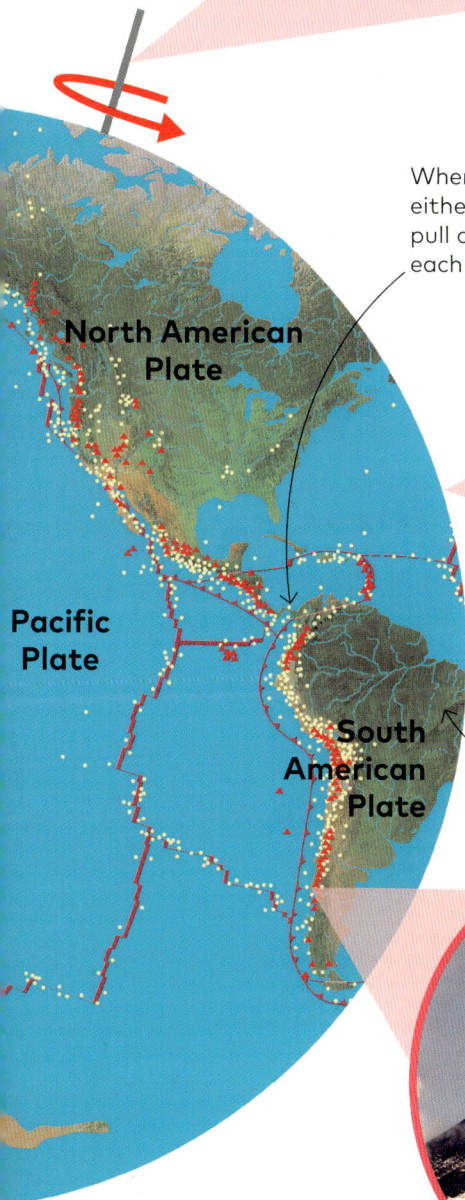

Axis of rotation

Tilt of axis varies during cycle

See pages 132–133 for the answers

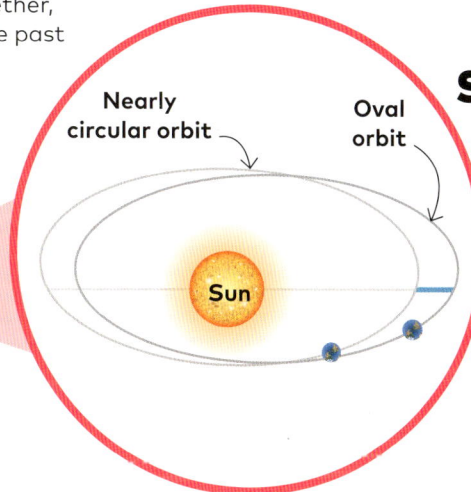

QUICK QUIZ

1. How can volcanic gases affect Earth's climate?

2. Is natural climate change a fast or slow process?

3. What are the rocky plates of Earth's crust called?

Where plates meet, they either bump together, pull apart, or slide past each other.

North American Plate

Pacific Plate

South American Plate

Plates carry continents over Earth's surface.

Nearly circular orbit

Oval orbit

Sun

Shifting orbit

Earth's orbit changes from oval-shaped to nearly circular over about 100,000 years. This varies the amount of energy the planet gets from the Sun.

How fast is natural climate change?

Natural changes to Earth's climate happen very slowly, usually over thousands or millions of years. The problem with today's human-caused global warming is that it is happening so fast. Animals and plants will have no time to adapt to the change.

Erupting volcanoes

Gases from volcanic eruptions hang in the atmosphere for a long time. Some volcanic gases can cool the climate, others can raise global temperatures.

When was the last ice age?

The "last" ice age started over 2.5 million years ago, and it hasn't ended yet! During an ice age, Earth's climate isn't always freezing. There are long, cold periods called glacials. These are broken up by shorter, warm periods called interglacials. We are living in an interglacial period today.

Ice cover

Glaciers and ice sheets cover large parts of Earth's land surface during the glacial periods of an ice age, and there is more sea ice, too.

How much ice was there?

Ice cover around the North Pole 20,000 years ago

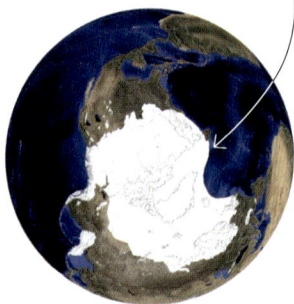

In the last glacial period, the ice cover was the greatest about 20,000 years ago. At that time, the world was probably about 5°C (10°F) colder. Much of North America and Northern Europe was under ice.

Reflecting heat

Ice and snow reflect a lot of the Sun's energy. More ice and snow at the start of an ice age means more heat is reflected and less is absorbed, so Earth gets even colder.

1. Why is Earth's ice melting faster today?

2. During an interglacial period, does ice cover grow or shrink?

3. When was Earth's ice cover last at its greatest?

See pages 132–133 for the answers

Growing and shrinking

As the climate flip-flops between glacial and interglacial periods, the ice cover grows and then shrinks. Today, Earth's ice is melting extra quickly because of global warming caused by human activity.

When will Earth turn cold again?

The warm interglacial period we're in now started over 11,000 years ago. It's possible that the next glacial period will begin in a few thousand years' time, and then Earth will turn cold again.

Glacial-interglacial cycles over the past 800,000 years

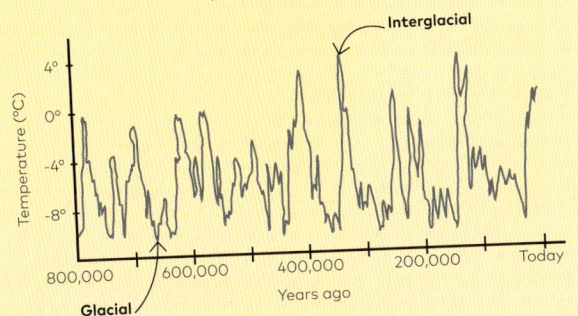

Temperature (°C)

Interglacial

4°
0°
-4°
-8°

800,000   600,000   400,000   200,000   Today

Years ago

Glacial

Has the climate ever been warmer than today?

Yes, many times, and sometimes much warmer. One such time was the Mesozoic Era, 252 to 66 million years ago, when dinosaurs roamed the Earth. The whole world was several degrees warmer than now, and there was less of a difference in temperature between the Equator and the poles.

Humid Earth

The start of the Mesozoic Era was hot and dry. It later became more humid, much like in the tropical regions around the Equator today.

Have the North and South poles always been frozen?

No. At different times in Earth's history, the poles have been ice-free. In the Mesozoic, there was probably little or no ice at the poles. Fossil finds show that in the late Mesozoic, dinosaurs and plants lived in Antarctica.

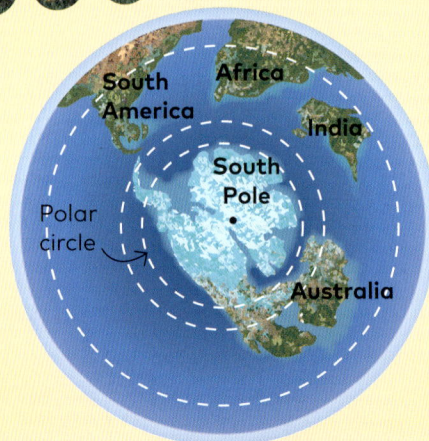

South America

Africa

India

South Pole

Polar circle

Australia

About 95 million years ago

Sago palm

Greenhouse plants

Plants thrived in greenhouse-like conditions, and new types of plant evolved. Conifers, ferns, and cycads (palm-like plants including the sago palm) were widespread. Flowering plants appeared in the late Mesozoic.

When was the climate last warmer than today?

Around 125,000 years ago, during the Eemian, the last interglacial period, Earth's climate was warmer. Ice sheets were smaller, and sea levels were higher. It was warm enough for elephants and hippos to live in Europe.

Reptile rulers

The Mesozoic is known as the Age of Reptiles. The warm climate was perfect for reptiles, some of which grew to huge sizes. Dinosaurs dominated life on land, and marine reptiles cruised the ocean.

TRUE OR FALSE?

1. The Mesozoic Era was the last time the climate was warmer than it is today.

2. The polar regions haven't always been ice-covered.

3. The Mesozoic is known as the Age of Amphibians.

See pages 132–133 for the answers

Why is Earth warming up so fast?

Earth's average temperature has risen by over 1°C (about 2°F) in the last 150 years. This is faster than we would expect the climate to change naturally. The rise is mainly caused by burning fossil fuels (coal, oil, and natural gas), which add carbon dioxide, methane, and other greenhouse gases to the atmosphere.

Generating energy

As Earth's population has grown, more energy is needed to power towns and cities. Generating energy at power stations often involves emitting greenhouse gases.

Deforestation

Trees absorb (take in) carbon dioxide from the air. Deforestation – cutting down forests – leaves fewer trees to absorb carbon dioxide, so this gas builds up in the atmosphere.

EVERY SECOND ACROSS THE WORLD, **MORE THAN 1 MILLION KG (2.2 BILLION LB) OF** CARBON DIOXIDE IS EMITTED.

Transport

Most forms of transport are powered by fossil fuels. The more cars, lorries, planes, and ships we make, and the more we use them, the more greenhouse gases are released into the atmosphere.

Landfills

As waste and rubbish rots and breaks down on landfill sites, it gives off methane and carbon dioxide. Humans make a lot of waste, so huge amounts of these gases are released.

TRUE OR FALSE?

1. Earth's temperature has stayed the same over the last 150 years.

2. Farm animals burp and fart carbon dioxide gas.

3. Our daily activities cause carbon dioxide to be released into the atmosphere.

See pages 132–133 for the answers

What is your carbon footprint?

Almost everything we do results in carbon dioxide and other greenhouse gases being released into the air – from turning on a light to buying food and driving a car. Your carbon footprint is the total amount of greenhouse gases your day-to-day activities produce.

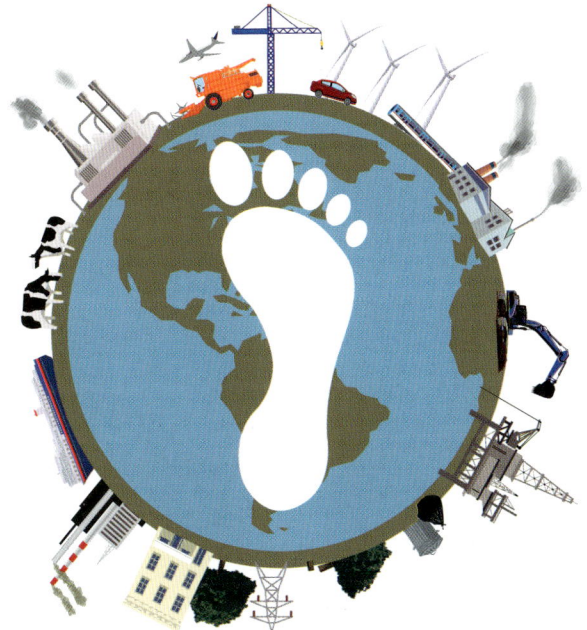

Farming

Some greenhouse gases are produced by growing crops, but more come from rearing animals for meat and dairy foods. This is because farm animals burp and fart lots of methane gas.

Polar crisis

The Arctic and Antarctic are warming faster than anywhere else on Earth, and the sea ice on the polar oceans is shrinking. Animals that rely on sea ice, such as polar bears, seals, and penguins, are losing their natural habitat as the ice dwindles.

Sinking islands

As oceans warm up, they expand and the sea level rises. Melting glaciers and ice sheets add water to the ocean, too. Rising sea levels are threatening low-lying islands such as the Maldives in Southeast Asia – and all the people and wildlife that live there.

How is global warming affecting our oceans?

Climate change is a challenge facing the world's oceans and everything living in them. Increasing temperatures and shrinking ice are resulting in the loss of habitats. Rising sea levels are a threat to islands and coastal areas, and important ocean currents are faced with disruption.

SCIENTISTS THINK ABOUT 80 PER CENT OF THE MALDIVES MAY BE UNDERWATER BY 2050.

Food chain collapse

Tiny organisms called phytoplankton are at the bottom of the marine food chain. If the ocean warms, they may not grow so well. Smaller plankton will mean less energy for the animals eating them – and also for their predators.

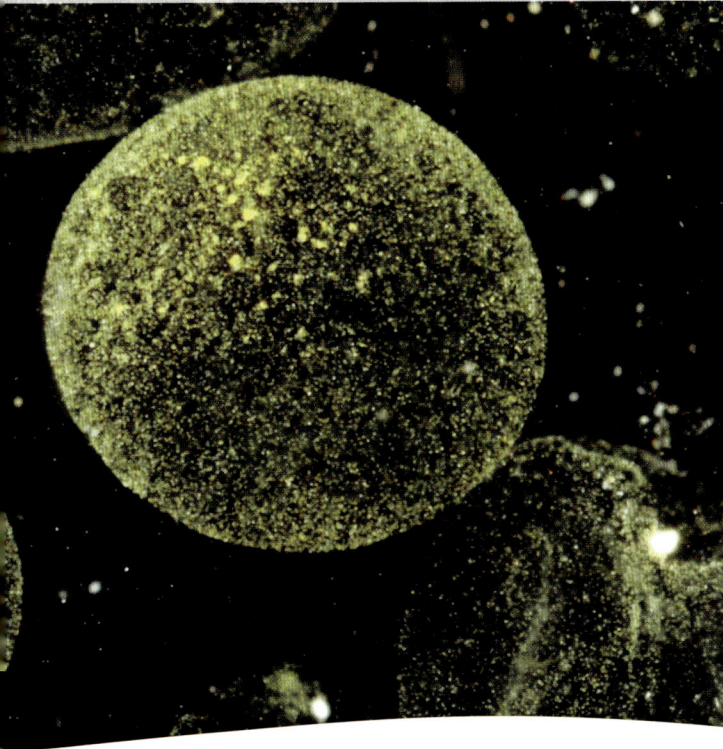

Coral bleaching

Coral reefs are bustling marine habitats made by tiny animals called polyps. Colourful algae live inside the polyps and provide them with food. If the water becomes too warm, the polyps spit out the algae. These animals become weak, white, and can even die.

Will global warming change our ocean currents?

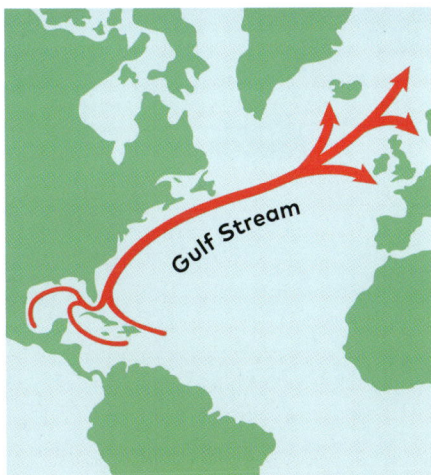

Ocean currents are vital in spreading heat energy around the planet, and they help regulate our weather and climate. Higher sea temperatures could disrupt the flow of major currents such as the Gulf Stream. Flowing in the North Atlantic, the Gulf Stream brings warm, tropical waters northwards. Without it, the climate of northwest Europe would be much colder.

Gulf Stream

QUICK QUIZ

1. What percentage of the Maldives is predicted to be underwater by 2050?

2. What do ocean currents regulate?

3. What gives colour to the coral reefs?

4. Which organisms are at the base of the marine food chain?

See pages 132–133 for the answers

How will global warming affect the weather?

Climate change is already changing the weather, and it will continue to do so as temperatures rise. Some places are becoming drier, others wetter. Around the world, the number of extreme weather events, such as heatwaves, droughts, and floods, is increasing, and storms are growing more intense.

DUE TO CLIMATE CHANGE, THE USA MAY HAVE **30 PER CENT MORE LIGHTNING** BY 2100.

Hotter temperatures

Heatwaves are becoming more common across different areas of the planet. As the climate continues to warm, summer temperatures are likely to get even hotter.

Worse droughts

In some parts of the world, droughts are getting worse as there is less rain. Soaring temperatures are drying up rivers, lakes, and the soil. Farm crops wither and die, threatening the food supply.

Will global warming affect the seasons?

Rising temperatures are causing spring to start earlier. This disrupts entire ecosystems of plants and animals because it is out of step with the natural rhythm of their lives.

TRUE OR FALSE?

1. Droughts can cause problems for farming.

2. A warmer atmosphere can hold more water.

3. Global warming is causing spring to start later.

See pages 132–133 for the answers

More floods

Hotter air can hold more water, which it draws up from warmer seas. When this moisture-laden air moves over land, it brings heavier rains that cause floods.

Fiercer storms

Higher ocean temperatures are making hurricanes and typhoons more intense. These storms gain more energy from the warmer ocean water, causing their winds to blow even stronger. Sea-level rise on top of larger storm surges will increase floods, too.

Countries are aiming to reduce carbon emissions to net zero by 2050.

Ditch fossil fuels

Using less coal, oil, and natural gas, and switching to clean energy sources, would reduce the greenhouse gases we emit.

Use alternative fuels

Unlike petrol and diesel cars, electric vehicles don't produce greenhouse gases when we drive them. Other fuels, such as hydrogen power, are also less harmful to the planet.

Electric car chargers are now common around the world.

How can the world reduce global warming?

We can't stop climate change overnight, but we can slow it down and stabilize it if we reduce greenhouse gas emissions on a global scale. This means people, businesses, and countries around the world all working together.

What is the Paris Agreement?

In 2015, representatives of many countries met in Paris to discuss the climate crisis. Together, they signed the Paris Agreement, a treaty with the shared goal of keeping the global temperature increase well below 2°C (3.6°F).

Reduce dairy and meat

Eating less dairy and meat will decrease the demand for animal farming. And if fewer animals are farmed, there will be lower methane emissions.

Rainforests cover around 6 per cent of land on Earth.

Protect nature

Green plants and seaweeds extract carbon dioxide from the atmosphere. We need to protect woodland and green spaces on land, and seagrass meadows and kelp forests in the ocean, too.

Campaign for change

For lasting change that makes a big difference, governments and businesses need to act. Campaigning can persuade governments to pass laws to protect the environment, and businesses to work in a more planet-friendly way.

QUICK QUIZ

1. What type of car fuel does not emit carbon dioxide?
   a) Petrol
   b) Electricity
   c) Diesel

2. When was the Paris Agreement signed?
   a) 1840
   b) 2015
   c) 2024

See pages 132–133 for the answers

Wind farms – clusters of wind turbines – are often built off the coast because sea winds tend to be stronger than those over land.

Wind power

A wind turbine uses a set of spinning, propeller-like blades called a turbine to harness the wind's energy and generate electricity. The biggest turbines can power thousands of homes.

What is clean energy?

Clean energy doesn't pollute the air or release greenhouse gases when it is produced. We can get clean energy from wind, flowing water, sunlight, and underground rocks. As well as helping us fight climate change, these natural energy sources won't run out.

THE COUNTRY OF ICELAND IS POWERED BY MORE THAN **80 PER CENT CLEAN ENERGY.**

Hydroelectricity

Huge amounts of water are stored behind a dam. When the water is allowed to fall through tunnels in the dam wall, it flows with great force and spins turbines. As the turbines spin, they drive electricity generators.

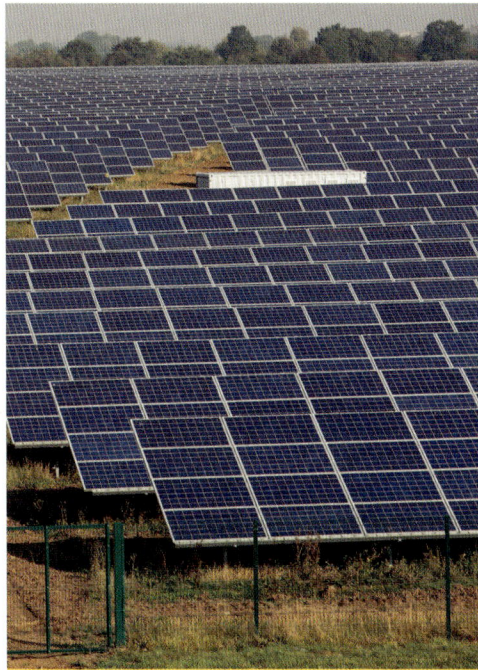

Solar power

A solar cell is a device that turns the energy of sunlight into electricity. Panels made up of solar cells can be fitted to roofs to provide power for buildings. A large group of solar panels on the ground is called a solar farm.

Geothermal energy

Earth's interior is hot. By drilling down into the rocks of Earth's crust, we can make use of this heat. Geothermal power stations pump up the hot water and turn it into steam. This drives turbines linked to generators.

How does a wind turbine work?

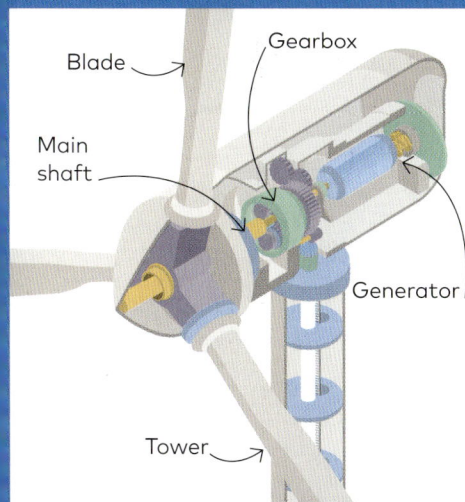

Blade
Gearbox
Main shaft
Generator
Tower

The propeller blades are mounted on a shaft linked to a gearbox. Cogwheels in the gearbox turn a second shaft connected to a generator. When the wind blows, all these parts turn quickly together. Inside the generator, a metal coil produces electricity as it spins within a magnetic field. In strong winds, any extra energy produced can be stored in special batteries for later use.

TRUE OR FALSE?

1. Hydroelectric power comes from the Sun.

2. Clean energy produces no air pollution.

3. We will one day run out of clean energy sources.

4. A wind turbine uses wind to make electricity.

See pages 132–133 for the answers

How can I help the planet?

Climate change needs to be tackled across the whole world, but that doesn't mean you can't do your bit to help! Doing simple, easy things like recycling your rubbish and using less electricity can make a big difference.

Plant more trees

Trees absorb the greenhouse gas carbon dioxide from the air, and release oxygen that we need to breathe. Deforestation is a global problem, so we need to plant lots of new trees to help keep our air healthy.

What are food miles?

Food miles describe the distance food has to travel to reach your plate from where it was produced. Food from far away will have been carried on planes, ships, or trucks that pollute the air. Try to buy food produced locally, which has travelled fewer food miles.

Bikes don't pollute the air like most cars do.

Use cars less often

The most planet-friendly ways to travel are walking and cycling. Using trains, buses, and trams is usually better than going by car, since most cars produce more greenhouse gases per passenger.

Recycle

Producing materials takes up a lot of energy. Recycling an object means we can use the materials it's made of to make something new. This can save energy, and each recycled item is one less item in landfill.

QUICK QUIZ

1. What do food miles measure?

2. What do trees absorb?

3. Which ways of travel are planet-friendly?

4. What can we switch to instead of meat and dairy?

See pages 132–133 for the answers

Packaging such as paper and cardboard, metal cans, and plastic and glass bottles can go into special recycling bins.

Think about your food

By eating more vegetables and plant-based foods, and less meat and dairy, you can reduce the size of your carbon footprint. If you have a garden or yard, try growing your own fruit and vegetables.

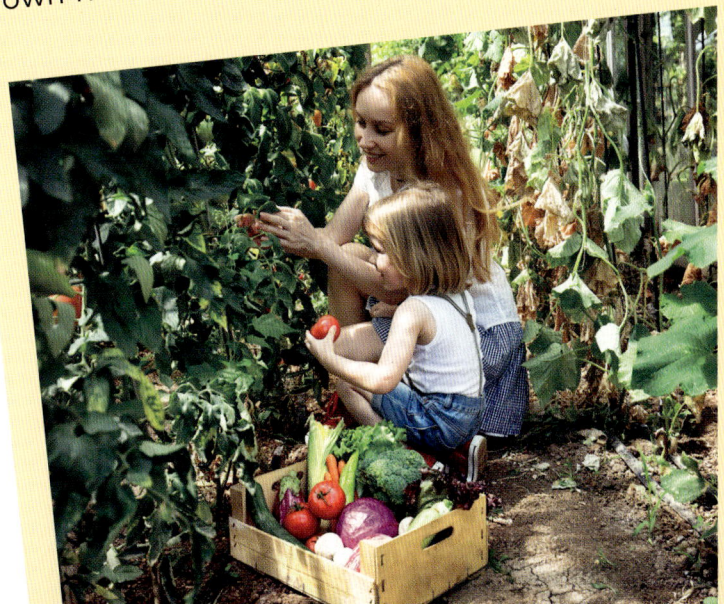

Save energy

Switch off things like lights and electrical devices when they're not in use, to save energy. This means power stations don't have to make as much electricity, so fewer greenhouse gases are produced.

Answers

Page 9
1) False. Air is mostly nitrogen.
2) True. 3) True. 4) False. Warm air rises and cool air sinks.

Page 11
1) c. 2) a.

Page 13
1) About 8 minutes. 2) Plants need sunlight to make their food. 3) The Sun will swell up and become a red giant star.

Page 15
1) True. 2) True. 3) False. Burning fossil fuels adds extra greenhouse gases to the atmosphere.

Page 17
1) Ocean-dwelling microbes called cyanobacteria. 2) The gases that made up the early atmosphere came from volcanoes. 3) 200 million years.

Page 19
1) False. Throughout the year, different parts of the Earth receive more sunlight as they point towards the Sun, then less sunlight as they point away from it. 2) False. Earth's closest point to the Sun is the perihelion. 3) True.

Page 21
1) Warm air. 2) Precipitation.
3) Cooler.

Page 23
1) c. 2) a.

Page 25
1) Auroras. 2) The Sun's corona, its hot, outer atmosphere.

Page 29
1) Contrails. 2) Miniscule, floating particles, such as dust, pollen, or grains of sea salt. 3) A cumulus cloud.

Page 31
Despite its flying-saucer shape, this is a cloud. Clouds like this are called lenticular clouds, which means they are lens shaped. They form where moist winds blow over mountains – and they are sometimes mistaken for UFOs!

Page 33
b.

Page 35
1) False. Ground-level clouds are called fog when you can see less than 1,000 m (3,000 ft) in front of you. 2) True. 3) False. Haze is caused by tiny particles in the air, such as dust, but not by water droplets.

Page 37
1) a. 2) c.

Page 39
1) c. 2) a.

Page 41
1) Lightning can heat air to temperatures up to 30,000°C (50,000°F). 2) Most lightning happens inside of clouds, and less often between the cloud and ground. 3) Ice crystals and freezing water droplets.

Page 43
1) c. 2) a.

Page 45
1) True. 2) True. 3) False. By day, cool air from the valley flows up the slopes.

Page 46
1) To help sailors judge wind strengths at sea. 2) 13 named levels.

Page 49
1) Six. 2) Because snow reflects the sunlight that falls on it back to our eyes, and since sunlight looks white, the snow does, too. 3) Powdery snow.

Page 51
Icicles are ice. They form when liquid water droplets freeze as they drip downwards.

Page 53
1) False. Clouds form when there is low pressure. 2) False. High pressure brings sunny days, but the weather may still be cold in winter. 3) True.

Page 54
1) Cold front. 2) Isobars.

Page 57
1) Cirrus clouds 2) Water vapour condenses around airborne spores to form water droplets and clouds, causing rain. 3) Algae.

Page 61
1) Permanently frozen ground beneath the soil in some polar areas. 2) Subtropical climate zones. 3) Four.

Page 63
1) b. 2) c.

Page 65
1) No. Nights can be very cold in hot deserts. 2) The Atacama Desert. 3) Because Antarctica's cold air can't hold much moisture, so there is very little precipitation, most of which is snow.

Page 67
1) In summer. 2) The sea.
3) From land to sea.

Page 69
1) a. 2) b.

Page 71
1) True. 2) False. Mountaintops are colder than the surrounding lowlands. The air gets colder the higher you go. 3) False. Wind chill makes it feel colder than the actual air temperature.

Page 73
1) a. 2) b.

Page 77
1) False. The fastest winds blow around the eye, close to the storm's centre. 2) True. 3) False. Tropical cyclones can be called hurricanes, typhoons, or cyclones, depending on where in the world they happen.

Page 79
1) Tropical cyclone. 2) 27°C (80°F).

Page 81
b.

Page 83
1) High pressure. 2) More common.
3) Thunderstorms.

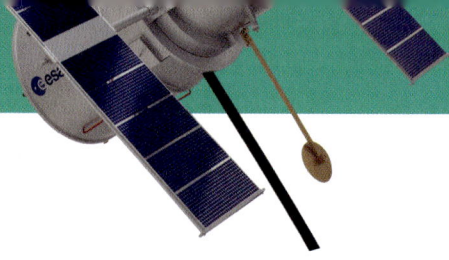

Page 85
1) Dry plant matter, such as leaves, grass, and branches. 2) When there is a long period of hot, dry weather and plenty of parched vegetation. 3) Because global warming is causing more heatwaves and lightning.

Page 87
1) False. Hail forms in cumulonimbus clouds. 2) True. 3) False. Some hailstones can also be much larger. 4) True.

Page 89
1) False. Sand grains are heavier than dust particles. 2) False. Dust storms form in dry regions with few plants and trees. 3) True.

Page 91
1) a. 2) b.

Page 93
1) b. 2) c.

Page 95
1) Snow squalls are common in winter months. 2) Snowploughs.

Page 99
1) 37°C (98.6°F). 2) The salt must be removed from the sea water. 3) Walls of dried mud keep homes from becoming too hot inside.

Page 101
Blubber.

Page 103
It's a desert plant called Welwitschia. The plant has only two leaves, each up to 8 m (26 ft) long, and covered in lots of pores to absorb water from fog and dew. The leaves wind round the plant, and split into straggly strips.

Page 105
1) a. 2) c.

Page 107
1) c. 2) a.

Page 109
1) About 160. 2) Radio waves. 3) It is fed into computers that use scientific equations to predict what the weather is likely to do.

Page 111
1) b. 2) a.

Page 115
1) Some volcanic gases can cool the climate, others can make it warmer. 2) Natural climate change happens very slowly. 3) Tectonic plates.

Page 117
1) Because of global warming caused by human activity. 2) Ice cover shrinks during an interglacial period. 3) About 20,000 years ago.

Page 119
1) False. The climate was warmer 125,000 years ago in the last interglacial period. 2) True. 3) False. The Mesozoic is known as the Age of Reptiles.

Page 121
1) False. Earth's average temperature has risen by over 1°C (about 2°F) in the last 150 years. 2) False. They burp and fart methane. 3) True.

Page 123
1) 80 per cent. 2) Ocean currents regulate our weather and climate. 3) Algae that live inside the coral polyps. 4) Phytoplankton.

Page 125
1) True. 2) True. 3) False. Global warming is causing spring to start earlier.

Page 127
1) b. 2) b.

Page 129
1) False. Hydroelectric power comes from flowing water. 2) True. 3) False. We will never run out of clean energy sources as they are renewable. 4) True.

Page 131
1) The distance food travels before reaching your plate. 2) Carbon dioxide. 3) Walking and cycling. 4) Plant-based food, fruit, and vegetables.

Quiz your friends!

Who knows the most about weather? Test your friends and family with these tricky questions. See pages 136–137 for the answers.

Questions

1. What is the name of this **type of cloud**?

6. Which is thinner, **MIST OR FOG**?

9. Which type of climate zone has **mild weather**?

3. What is the **huge storm** on Jupiter called?

5. How fast can really **big hailstones** fall?

2. What shape are the **ICE CRYSTALS** that form in clouds?

4. Why are polar bears **ENDANGERED**?

7. In which country are **tornadoes** most common?

8. What direction do tropical cyclones swirl in the **northern hemisphere**?

12. What is **CLEAN ENERGY**?

10. What do we call a **SPINNING COLUMN** of dust-filled air?

13. Why are **sea levels rising**?

11. Why do we see a **lightning flash** before we hear **thunder**?

14. What is the order of **COLOURS IN A RAINBOW**?

Answers

1. Cumulonimbus.

6. Mist.

9. Temperate climate zone.

12. Energy that doesn't release any pollution or greenhouse gases when it's used.

3. Great Red Spot.

2. Hexagonal (six-sided).

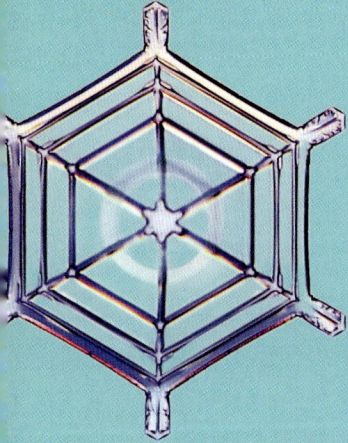

8. Anticlockwise.

5. The BIGGEST hailstones can fall at **160 kph (100 mph)!**

11. Because light travels **faster** than sound.

4. Because **global warming** is melting the sea ice on which they live.

7. The USA.

10. A dust devil.

13. Sea water is **EXPANDING** as it warms up, and **ice sheets and glaciers are melting,** and flowing into the ocean.

14. Red, orange, yellow, green, blue, indigo, violet.

Glossary

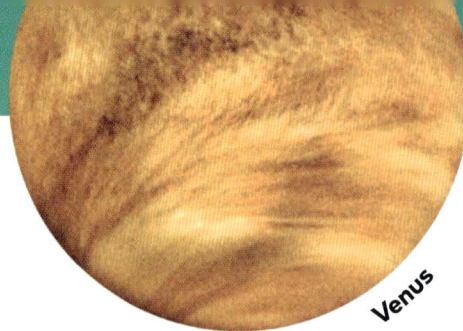

Venus

Mars

air mass
Large body of air – all with the same temperature, pressure, and humidity

air pressure
Weight of air pressing down on the ground or any other surface; also called atmospheric pressure

anemometer
Instrument for measuring wind speed

atmosphere
Blanket of gases surrounding Earth

aurora
Shimmering bands or swirls of coloured light in the night sky over regions near the poles

barometer
Instrument for measuring air pressure

Beaufort scale
A scale for estimating the strength of the wind, based on what you can see happening

black ice
A thin layer of hard-to-see ice that forms when rain or drizzle falls on a cold road

blizzard
A storm in which snow is blown into the air by very strong winds

campaigning
Taking action in order to bring about change

carbon footprint
The amount of carbon dioxide your daily activities cause to be released into the atmosphere every year

Aurora

cell
Tiny part of a living thing; bacteria are made up of just one cell, but humans have trillions of cells

cirrus
Wispy clouds of ice crystals that form high in the sky

clean energy
Energy that creates no greenhouse gas emissions when produced and used

climate
Normal weather pattern of a large area, averaged over a long period of time

climate change
Long-term change in Earth's weather patterns

climatologist
Scientist who studies weather patterns over long periods of time

clouds
Masses of water droplets and ice particles suspended in the air

cold front
Boundary between two air masses, where the cold air is moving towards the warm air in front of it

condensation
Transformation from a gas to a liquid

cumulonimbus
Huge cloud that produces heavy showers and thunderstorms

cumulus
Fluffy cloud with a flat bottom and rounded top

deforestation
The clearing of forest and woodland

desert
Place with very little rainfall, or no rain at all

dew
Moisture from the air that has condensed into water droplets on objects

downdraught
Downward flow of air

drizzle
Light rain made of drops smaller than 0.5 mm (0.02 in) across

drought
Extended period without rain, or much less rain than normal

dust storm
Thick, rolling cloud of dust whipped up by the wind

Equator
The line around Earth's middle, halfway between the poles

evaporation
The change of a liquid to a gas

fog
Cloud of tiny water droplets near the ground, through which you can see less than 1,000 m (3,000 ft)

food miles
The distance food travels before it reaches your plate

fossil fuels
Fuels formed from the buried remains of living things; coal, oil, and gas are fossil fuels

front
Boundary between two air masses

frost
White ice crystals that form on cold surfaces when moisture in the air freezes

generator
Machine that produces electricity

geothermal power
Electricity generated by using heat from deep below Earth's surface

glacier
Large "river" of slowly moving ice

global warming
Long-term increase in the temperature of Earth's surface, mainly caused by humans burning fossil fuels

greenhouse effect
Trapping of the Sun's heat by gases in Earth's lower atmosphere

greenhouse gases
Gases in the atmosphere that trap heat; carbon dioxide, methane, and water vapour are some of the main greenhouse gases

habitat
Natural home of a living thing

hail
Pellets of ice that fall from clouds

haze
Cloud of tiny, dry particles suspended in the air that makes far-off objects look fuzzy on warm days

hemisphere
Half of Earth; the Equator divides Earth into northern and southern hemispheres

humidity
Measure of water vapour in the air

hurricane
Tropical cyclone in the North Atlantic

hydroelectricity
Electricity produced when falling running water from a dam turns a turbine linked to a generator

hygrometer
Instrument used for measuring humidity

ice age
Cold period in Earth's history, when ice sheets are much larger

ice sheet
Vast layer of ice on land

interglacial
Warmer period during an ice age

isobar
Line on a weather map that joins places with the same air pressure

kelp forest
Large seaweeds (algae) that form thick "forests" near the the shore in cool, shallow waters

landfill
Getting rid of rubbish by putting it in large holes in the ground

lightning
Sudden flow of electricity from a cloud

meteorologist
Scientist who studies the atmosphere to understand the weather

microclimate
Climate of a small area

migration
Movement of people or animals from one place to another

Saturn

mist
Thin cloud near the ground

monsoon
Wind that brings alternate wet and dry seasons at the same time each year

paleoclimatologist
Scientist who studies what Earth's climate was like long ago

photosynthesis
Process in which plants, algae, and some bacteria use energy from sunlight to make food

power station
Place where electricity is produced

precipitation
All forms of water that fall to the ground, such as rain, snow, hail, dew, fog, and mist

prevailing wind
Main direction the wind blows from in a certain place

psychrometer
Type of hygrometer; instrument for measuring humidity

radar
Method of detecting and tracking an object by bouncing radio waves off it

rain gauge
Instrument used to collect and measure rainfall

rain shadow
Area of lower rainfall on the sheltered side of a mountain

rainbow
Arc of colours that appears in the sky when sunlight passes through raindrops

recycling
Reusing waste items, or using the materials they contain to make new items

renewable energy
Energy from a source that does not use up Earth's resources, and does not run out

sandstorm
Cloud of wind-blown sand grains that moves over the desert

season
One of the main weather periods of each year

smog
Originally fog mixed with smoke, but now usually a haze that forms in polluted air in strong sunshine

snow
Ice crystals that fall from clouds and stick together to form snowflakes

solar cell
Device that converts sunlight directly into electricity

solar panel
Sheet of solar cells that produces electricity from sunlight

solar wind
Stream of tiny, electrically charged particles that constantly flows out from the Sun into space

space weather
Bursts of radiation, particles, and magnetism from the Sun

storm
Strong winds, between gale and hurricane force

stratus
Sheet-like clouds that sometimes cover the whole sky

supercell
Vast thunderstorm cloud that contains a spinning updraught

sustainable transport
Transport that produces little or no air pollution, or harm to the environment

tectonic plate
One of the pieces that make up Earth's rocky crust

temperate climate
Mild climate; places with a temperate climate lie between the warm tropics and the cold polar regions

thunder
Sound made by expanding air during a flash of lightning

tornado
Spiral of air spinning at high speed over land

transpiration
Process in which plants take up water from the soil with their roots and release it into the air from their leaves as water vapour

tropical cyclone
Very powerful storm that brews up over warm ocean water in tropical regions

tropics
Regions either side of the Equator, which are warm all year

troposphere
Lowest or innermost layer of Earth's atmosphere, where most of the weather takes place

typhoon
Tropical cyclone that forms over the Pacific Ocean

updraught
Upward flow of air

warm front
Boundary between two air masses, where the air behind the front is warmer than the air ahead of it

water vapour
Water in the form of a gas

waterspout
Column of rapidly spinning air over water

weather
Conditions in the atmosphere at a particular time and place

wildfire
Fire in a natural area, such as forest or grassland

wind
Stream of air moving from one place to another

wind turbine
Set of spinning blades, turned by the wind, that drives an electricity generator

windchill
The chilling effect of wind on your body – it takes heat away from your body as it blows over you

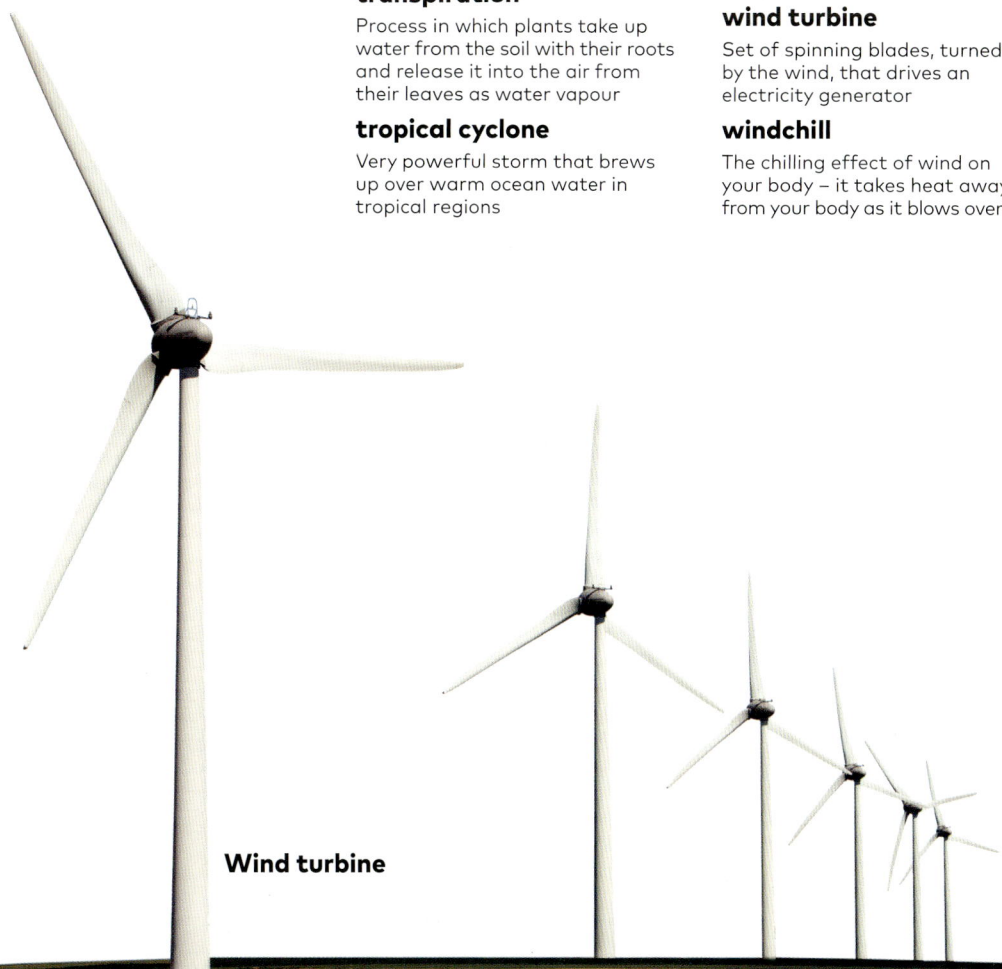

Wind turbine

Index

Use these pages to help you find what you're looking for in the book.

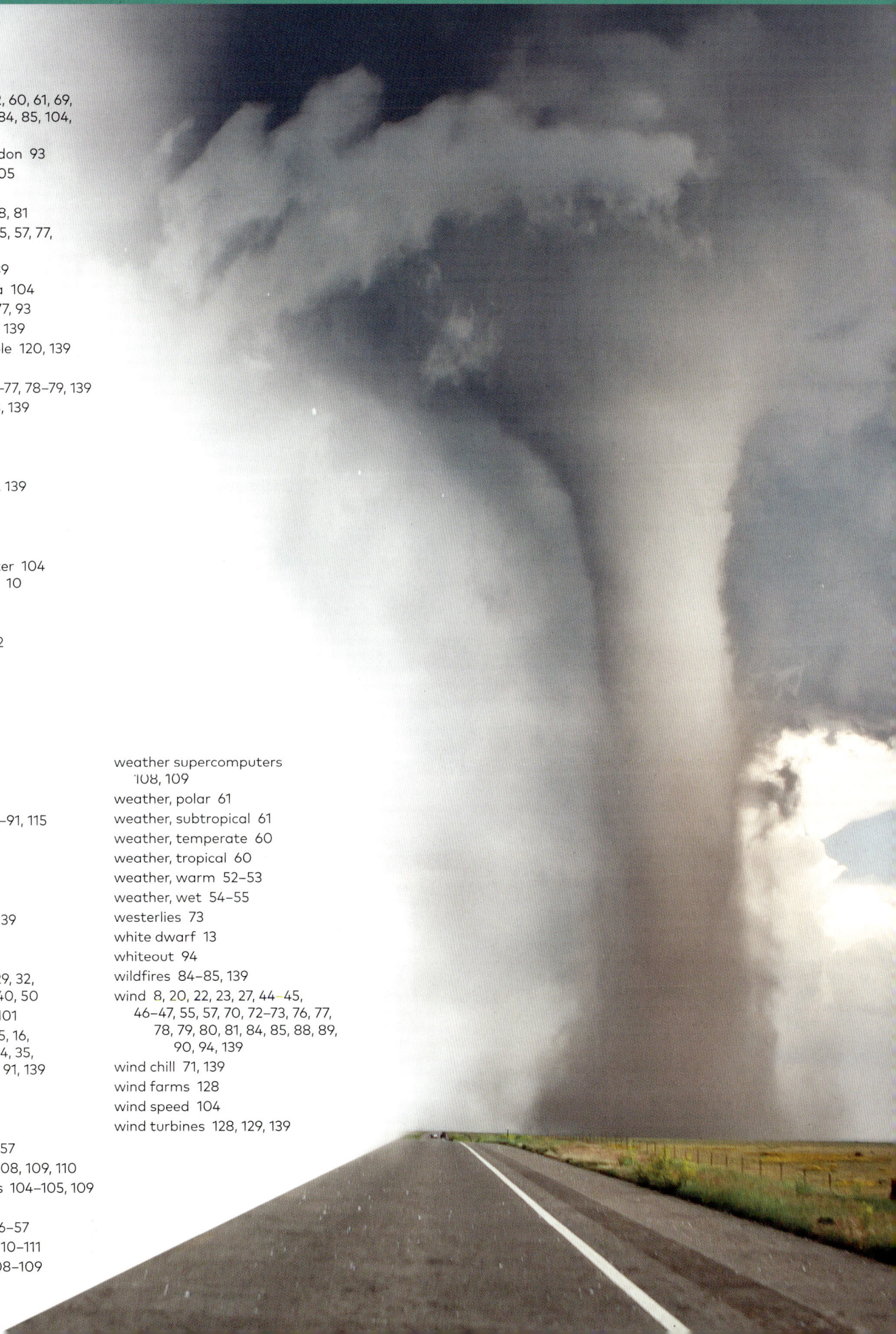

Acknowledgements

THIS EDITION

Dorling Kindersley would like to thank Sakshi Saluja and Samrajkumar S for picture library assistance, Polly Goodman for proofreading, and Claire Sipi for indexing.

The publisher would like to thank the following for their kind permission to reproduce their photographs:
(Key: a-above; b-below/bottom; c-centre; f-far; l-left; r-right; t-top)

1 123RF.com: Alhovik (tl/Tornado). **Dreamstime.com:** Connect1 (tl/Barometer); Constantin Opris (tc); Wenani (c); Torian Dixon / Mrincredible (c/Jupiter). **ESA:** AOES Medialab (ca). **Getty Images / iStock:** Imamember (tl). **NASA:** JPL / USGS (tr). **Science Photo Library:** NASA (bc). **US National Science Foundation:** Kenneth Libbrecht, Caltech (cr/X2). **2 123RF.com:** Ufuk Zivana (cr). **Dreamstime.com:** Sergey Novikov (cl). **Getty Images:** 500px / Josh Heidebrecht (c). **2–3 Dreamstime.com:** Bigjohn3650 (c). **4–5 Shutterstock.com:** Farjana Rahman (bc). **4 Science Photo Library:** Kenneth Libbrecht (cra). **5 123RF.com:** Irochka (tc). **Dreamstime.com:** Jan Martin Will (tr). **Science Museum Group:** (cb). **6 Dreamstime.com:** Elena Schweitzer (tl). **6–7 Getty Images:** 500px / GRobert (bc). **7 Alamy Stock Photo:** Steve Bloom Images / Nick Garbutt (t). **9 Dreamstime.com:** Rosa Frei (tr). **10 Getty Images / iStock:** Guvendemir (bl). **PunchStock:** Westend61 / Rainer Dittrich (br). **12 Dreamstime.com:** Elena Schweitzer (tl). **Shutterstock.com:** Sander van der Werf (tr). **13 Getty Images / iStock:** E+ / Cinoby (tl). **15 Shutterstock.com:** DimaBerlin (tl). **16 Getty Images:** 500px / GRobert (crb); 500px / Josh Heidebrecht (bl). **17 Alamy Stock Photo:** Blickwinkel / F. Fox (bl); Steve Bloom Images / Nick Garbutt (crb). **19 Alamy Stock Photo:** Science History Images (bl). **20 Dreamstime.com:** Sergey Novikov (tr); Satjawat Boontanataweepol (tl). **Getty Images / iStock:** E+ / 35007 (c). **21 Alamy Stock Photo:** Art Directors & TRIP / Helene Rogers (bl); Steve Bloom Images / Nick Garbutt (tl). **Dreamstime.com:** Lehmanphotos (tr); Serhii Yevdokymov (tc). **22 ESA:** NASA (clb). **NASA:** Johns Hopkins University Applied Physics Laboratory / Carnegie Institution of Washington. (cr). **23 Alamy Stock Photo:** Sipa USA (tl). **Dreamstime.com:** Nerthuz (bl). **NASA:** JPL / USGS (cra); JPL-Caltech / SSI (cr). **25 Dreamstime.com:** Dgmate (tc). **NASA:** SDO (bc). **26 Dreamstime.com:** Constantin Opris (cl). **Shutterstock.com:** Mihai Simonia (b). **26–27 Steve Setford:** (t). **27 Getty Images / iStock:** Vychegzhanina (tr). **29 Dreamstime.com:** Russ Heinl (bl). **30 Alamy Stock Photo:** Tim Gainey (bl); Genevieve Vallee (cra). **Dreamstime.com:** Danflcreativo (br). **Getty Images:** Moment / Supachai Panyaviwat (cr). **31 123RF.com:** Kvitkanastroyu (cla); Sergeymakarenko (bc). **Alamy Stock Photo:** Daybreak Imagery (r); imageBROKER.com GmbH & Co. KG / Frank Schneider (cl). **Dreamstime.com:** Constantin Opris (bl); Michael Piepgras (tl); Miguel Perfectti (tr). **32 123RF.com:** Yemelyanov (tr). **Dreamstime.com:** Minnystock (br). **Shutterstock.com:** Anurak Pongpatimet (bl). **33 Depositphotos Inc:** Dailajphoto (tl). **Dreamstime.com:** Retouch Man (crb). **Getty Images / iStock:** Vychegzhanina (bl). **34 Getty Images / iStock:** E+ / Deejpilot (Background). **Getty Images:** Hulton Archive / Keystone (clb). **35 Alamy Stock Photo:** Amer Ghazzal (bl). **Dreamstime.com:** Jmrocek (Background). **36 Alamy Stock Photo:** imageBROKER.com GmbH & Co. KG / Kevin Prönnecke (Background). **37 Depositphotos Inc:** Pavsie (Background). **38–39 Getty Images:** Moment / Mikroman6. **39 Getty Images:** Cavan Images / Julia Crim (cb). **Shutterstock.com:** Muratart (tl). **40–41 Shutterstock.com:** Mihai Simonia. **40 Getty Images:** John Finney Photography (bl). **41 Alamy Stock Photo:** Zone3 (tr). **42 Depositphotos Inc:** EpicStockMedia (bl). **43 Dreamstime.com:** Philip Steury (br). **44 Dreamstime.com:** Serdar Corbac (tr); VectorMine (bl). **44–45 Dreamstime.com:** Vacclav (Background). **45 Dreamstime.com:** Serdar Corbac (tl). **46 Alamy Stock Photo:** IanDagnall Computing (cl). **48–49 Dreamstime.com:** Ekaterina Tomko (ash particles); Chumphon Whangchom (Background). **Science Photo Library:** Kenneth Libbrecht. **49 Alamy Stock Photo:** Thomas Biegalski (br). **Getty Images / iStock:** E+ / Imgorthand (bl); Image Source / Jakob Helbig (bc). **US National Science Foundation:** Kenneth Libbrecht, Caltech (tc). **50 Steve Setford:** (tr). **Shutterstock.com:** Edgaras Borotinskas (tl). **51 Alamy Stock Photo:** Mauritius Images GmbH / Martin Siepmann (tl); Graham Turner (bc); Buddy Mays (br). **Dreamstime.com:** Makkis (tr). **52 Depositphotos Inc:** Belikovand. **Dreamstime.com:** Luckypic (t). **53 Alamy Stock Photo:** imageBROKER.com GmbH & Co. KG / Frank Schneider (ca); Genevieve Vallee (t). **Depositphotos Inc:** Ale-ks. **Dreamstime.com:** G?nter Albers (bl). **Getty Images:** Moment / Supachai Panyaviwat (cb). **Shutterstock.com:** Creative Travel Projects (bc). **54 Alamy Stock Photo:** Manfred Gottschalk (tr). **56 Alamy Stock Photo:** Image Professionals GmbH / LOOK-foto (c). **Paul McLoughlin:** (cl). **Trevor Rickard:** (cr). **57 Alamy Stock Photo:** All Canada Photos / Ken Gillespie (cl). **Getty Images:** Photodisc / Donovan Reese (c). **naturepl.com:** Hiroya Minakuchi (cr). **Science Photo Library:** Eye of Science (tl). **58–59 Alamy Stock Photo:** Cavan Images (tc). **Getty Images:** Moment / Comezora (b). **59 Shutterstock.com:** Sparc (tr). **60 Alamy Stock Photo:** Cavan Images (cl). **Dreamstime.com:** Nalidsa Sukprasert (tl). **60–61 Alamy Stock Photo:** Dimitrios Karamitros (Map). **61 Dreamstime.com:** Dimmushu (tr); Bradley Hay (cr). **62–63 Shutterstock.com:** Sparc. **63 Alamy Stock Photo:** J.Enrique Molina (cla); Rosanne Tackaberry (cra). **64 Alamy Stock Photo:** Jan Wlodarczyk (Background). **Getty Images / iStock:** E+ / Cinoby (tr). **65 Dreamstime.com:** Marktucan (br). **White Desert.com. 66 Getty Images:** Moment / Mayur Kakade. **67 Dreamstime.com:** Dimaberkut (tl); Hecos255 (tc). **68 Shutterstock.com:** Scott Riley (bl). **68–69 Alamy Stock Photo:** Wolfgang Kaehler. **69 Science Photo Library:** Bernhard Edmaier (cra). **70–71 Getty Images:** Moment / Comezora. **71 Alamy Stock Photo:** Associated Press / Mel Evans (br). **73 Alamy Stock Photo:** Panther Media GmbH / Dagmara (bl). **74 Shutterstock.com:** Mikadun (tl). **74–75 Getty Images:** Moment / Willoughby Owen (b). **75 Getty Images:** Moment Open / John Crux Photography (t). **76 Getty Images:** Win McNamee (bl). **76–77 Getty Images / iStock:** E+ / FrankRamspott. **77 NASA:** (tc). **79 Alamy Stock Photo:** Science History Images / Photo Researchers (tl). **80 Alamy Stock Photo:** Ryan McGinnis (bl). **Dreamstime.com:** John Sirlin (cl). **Shutterstock.com:** Aramiu (clb). **80–81 Getty Images:** Moment / Willoughby Owen. **81 Science Photo Library:** Roger Hill (tr). **82 Dreamstime.com:** Roman Slavik (bl). **83 Alamy Stock Photo:** Julian Eales (br). **84 Getty Images / iStock:** M_Pavlov (tl). **84–85 Getty Images:** Moment Open / John Crux Photography (t). **85 Alamy Stock Photo:** Australian Associated Press / David Mariuz (br). **Getty Images / iStock:** The Image Bank / Jupiterimages (clb). **86 Dreamstime.com:** Chaiyon021 (Background). **Science Photo Library:** Jim Reed. **87 Depositphotos Inc:** Adam_R (tr). Joshua Soderholm (Bureau of Meteorology): (tc). **88–89 Alamy Stock Photo:** JordiStock. **88 Getty Images / iStock:** PamelaPeters (br). **89 Dreamstime.com:** Kelly Vandellen (br). **90 Alamy Stock Photo:** Arctic Images / Ragnar Th Sigurdsson (bl). **90–91 Alamy Stock Photo:** Nature Picture Library / Erlend Haarberg. **91 Alamy Stock Photo:** Stefano Zaccaria (crb). **93 Alamy Stock Photo:** Robertharding / Alex Treadway (br). **Reuters:** Jorge Adorno. **94–95 Shutterstock.com:** Mikadun. **94 Ray Boren/Earth Science Picture of the Day:** Ray Boren / Earth Science Picture of the Day (bl). **95 Alamy Stock Photo:** imageBROKER.com GmbH & Co. KG / Guenter Fischer (tr). **Getty Images:** Corbis Historical / Christopher Morris (bl). **96 Shutterstock.com:** Jonathan Manjeot. **96–97 Dreamstime.com:** Valio84sl (tc). **97 ESA:** (ca). **98 Getty Images:** Moment / Posnov (c). **James H Morris:** (bl). **Anthony Ham:** (br). **99 Johnny Haglund:** (c). **Shutterstock.com:** Alexandra Kovaleva (tc). **100 Alamy Stock Photo:** Nature Picture Library / Steven Kazlowski (tr). **Dreamstime.com:** Alicenerr (tl/Background). **Getty Images / iStock:** FredRood (tl). **101 Alamy Stock Photo:** Minden Pictures / D. Parer & E. Parer-Cook (tl). **Dreamstime.com:** Bernard Breton (br). **Science Photo Library:** Power And Syred (bl). **Shutterstock.com:** Farjana Rahman (tr). **102 Getty Images / iStock:** Aimintang (bl). **Shutterstock.com:** Jonathan Manjeot. **103 123RF.com:** Normankrauss (tr). **Alamy Stock Photo:** Brian & Sophia Fuller (bc). **Science Photo Library:** Duncan Shaw. **104 Alamy Stock Photo:** Susan Norwood (tl); P Tomlins (tr). **Depositphotos Inc:** Serezniy (tc). **105 Alamy Stock Photo:** imac (tc). **Getty Images:** De Agostini Picture Library (cr). **S. Brannan & Sons Ltd:** S. Brannan & Sons Ltd (tl). **Science Museum Group:** (tr). **Science Photo Library:** Paul Rapson (bl). **106 Alamy Stock Photo:** Laura Clay Ballard (c); Alison Thompson (tl). **Dreamstime.com:** Valio84sl (tr). **106–107 Alamy Stock Photo:** Simon Colmer (bc). **107 Alamy Stock Photo:** Northern Nights Photography (tl). **Dreamstime.com:** Photographerlondon (br). **Getty Images / iStock:** bocero1977 (cr). **108–109 Dreamstime.com:** Uatp1 (tc). **ESA:** (bc). **108 Getty Images / iStock:** E+ / 4FR (cr). **109 Alamy Stock Photo:** Ben Welsh Premium (cr); Westend61 GmbH / Sebastian Dorn (cra). **Getty Images:** Kyodo News (tl). **NOAA:** (tr). **Science Photo Library:** Karim Agabi / Eurelios (bl). **110 Alamy Stock Photo:** Robertharding / Michael Nolan (tr); Torontonian (tl). **111 Alamy Stock Photo:** NASA Image Collection (tl); Science Photo Library / Mark Garlick (crb). **Getty Images / iStock:** E+ / Somethingway (cl). **112 Getty Images:** Stone / Paul Souders (tl). **112–113 Alamy Stock Photo:** Snorri Gunnarsson (bc). **113 Alamy Stock Photo:** Realimage (t). **114 Alamy Stock Photo:** Viktor Posnov (cb). **Shutterstock.com:** Mathias Berlin (tr). **114–115 Science Photo Library:** Gary Hincks (c). **115 Alamy Stock Photo:** Barrie Cooper (br); Francisco Negroni (bc). **116–117 Dreamstime.com:** Bennymarty. **116 NOAA:** (bl). **117 Dreamstime.com:** Luckypic (t). **118–119 Depositphotos Inc:** Allvisionn. **119 Alamy Stock Photo:** The Natural History Museum (cra). **Dreamstime.com:** Khunaspix (tl). **120 Dreamstime.com:** Vladvitek (cr). **120–121 Alamy Stock Photo:** Frank Bienewald (bc). **Getty Images / iStock:** E+ / ESKA (tc). **121 Alamy Stock Photo:** Wayne Hutchinson (bl). **Dreamstime.com:** K G (tl). **122 Getty Images:** Stone / Paul Souders (tl). **Shutterstock.com:** Tatiana Popova (tr). **123 Alamy Stock Photo:** imageBROKER.com GmbH & Co. KG / Steve Trewhella (tl). **Dreamstime.com:** OSweetNature (bl). **Getty Images:** Moment / Federica Grassi (tr). **124 Alamy Stock Photo:** USDA Photo (br). **Dreamstime.com:** Soumen Hazra (bl). **125 Alamy Stock Photo:** Mike Hill (br); Realimage (tl); Imaginechina Limited (bl). **126 Alamy Stock Photo:** Corine van Kapel (tl); Justin Kase z12z (cra). **126–127 Dreamstime.com:** Yuri Arcurs (bc). **127 Dreamstime.com:** Artushfoto (c); Matthew Dixon (tl); Saiko3p (tc). **128 Alamy Stock Photo:** Rob Arnold (t). **129 Alamy Stock Photo:** Ashley Cooper (tc); Snorri Gunnarsson (tr). **Shutterstock.com:** Evgeny_V (tl). **130 Alamy Stock Photo:** RichardBaker (bl). **Dreamstime.com:** Aurovenkatesh (tr); Pressmaster (br). **130–131 Dreamstime.com:** Vectorplus (c). **131 Getty Images / iStock:** E+ / Portishead1 (br); SouthWorks (bl). **Shutterstock.com:** Hurst Photo (tl). **133 Dreamstime.com:** Sergey Novikov. **134 Dreamstime.com:** Sergey Uryadnikov / Surz01 (cl). **134–135 Alamy Stock Photo:** Wolfgang Kaehler (Background). **Dreamstime.com:** Nalidsa Sukprasert. **135 Alamy Stock Photo:** Zone3 (cb). **Dreamstime.com:** Chaiyon021 (tr/ Background); Nerthuz (tc). **Science Photo Library:** Jim Reed (tr). **136–137 Alamy Stock Photo:** Wolfgang Kaehler (Background). **136 123RF.com:** Smileus (crb). **Dreamstime.com:** Jmrocek (cra); Jan Martin Will (cb). **137 Getty Images / iStock:** E+ / FrankRamspott (ca); PamelaPeters (cr). **Getty Images:** Stone / Paul Souders (tr). **Science Photo Library:** Kenneth Libbrecht (cra). **Shutterstock.com:** Pixssa (b); Tatiana Popova (clb). **138 Dreamstime.com:** Dgmate (bl). **ESA:** NASA (tc). **139 Dreamstime.com:** Delstudio (br). **NASA:** JPL / USGS (tl). **140–141 Depositphotos Inc:** Allvisionn (b). **142 Alamy Stock Photo:** Stefano Zaccaria (tl). **143 Getty Images:** Moment / Willoughby Owen. **144 Science Photo Library:** Karim Agabi / Eurelios (br)

Cover images: *Front*: **Dreamstime.com:** Connect1 tl, Constantin Opris tc, Wenani c; **ESA:** AOES Medialab ca; **Getty Images / iStock:** alhovik cla, Imamember tl/ (Rooster); **Science Photo Library:** NASA bc; *Back*: **Corbis:** Repro cl; **Dreamstime.com:** Constantin Opris cr, Positiveflash c; **Getty Images / iStock:** goinyk clb; **NASA:** JPL / USGS cra; **Science Photo Library:** Kenneth Libbrecht cla; *Spine*: **123RF.com:** Alhovik t